Modern theodolites and levels

Second edition

M. A. R. Cooper B.Sc., A.R.I.C.S.
The City University, London

GRANADA
London Toronto Sydney New York

Granada Publishing Limited – Technical Books Division
Frogmore, St Albans, Herts AL2 2NF
and
36 Golden Square, London W1R 4AH
866 United Nations Plaza, New York, NY 10017, USA
117 York Street, Sydney, NSW 2000, Australia
100 Skyway Avenue, Rexdale, Ontario, Canada M9W 3A6
61 Beach Road, Auckland, New Zealand

British Library Cataloguing in Publication Data
Cooper, M. A. R.
Modern theodolites and levels— 2nd ed.
1. Theodolites   2. Levels (Surveying instruments)
I. Title
526.9′028   TA575

ISBN 0-246-11502-5

First published in Great Britain 1971 by Crosby Lockwood & Son Ltd
Second edition 1982 by Granada Publishing Ltd

Typeset by V & M Graphics Ltd, Aylesbury, Bucks
Printed and bound in Great Britain by
William Clowes (Beccles) Limited, Beccles and London

Granada ®
Granada Publishing ®

To my parents

Contents

Preface to the second edition

A new edition of this textbook has been made necessary by the greatly increased use of electronics in surveying instruments during the decade since the original publication. This edition therefore includes details of the principles and applications of automatic optoelectronic scanning of coded circles and registration of data.

The opportunity has been taken to revise, rewrite and bring up to date information included in the first edition. The reader will find that the combination of new material, the revised and expanded original contents and a completely reset text have resulted in what is, in effect, a new book. However, the original aim and form of the presentation remain unaltered. The aim is to present the theory of theodolites and levels with an emphasis on the principles of instrument construction, sources of error and instrumental accuracies. The reason for such emphasis is that the author believes it is important to design a survey operation to meet a certain specification, particularly in large-scale surveys and setting out where tolerances are becoming smaller as costs increase. The first stage in a satisfactory selection of instruments and methods is an understanding of the errors in theodolites and levels and of the ways in which these errors can be eliminated or reduced. It is hoped that this book will assist in this. The form of the presentation of the material is explained in the Introduction.

The book is primarily intended to be used by students studying for first degrees and diplomas in surveying and civil engineering. Practitioners of land surveying and civil engineering will find the descriptions of the principles and applications of mechanical and electronic components useful in understanding the developments that have taken place since their formal studies ended. It is hoped that this understanding will lead to more efficient use of the newer instrumentation. It is also intended that this book will serve as an adequate text covering instrumentation for the measurement of angles and differences in levels by mechanical engineers in a workshop or laboratory. Students and practitioners of disciplines such as geography, building, architecture, archaeology and town planning which necessitate from time to time the use of theodolites and levels will find the book useful as a reference text covering modern instrumentation. Some hints to users about observing and recording are included for those whose knowledge of survey procedures is limited, but this book is not intended to be used as a source of information on surveying procedures and methods.

I have had to rely on several individuals and manufacturers for assistance in the collection of information and in preparing it for publication. I am particularly grateful to J. N. Hooker and K. D. Peak. Jim Hooker has produced those photographs not acknowledged elsewhere and has always responded with enthusiasm to my suggestions. Ken Peak has drawn all the new figures and revised some old ones, often on the basis of my almost incomprehensible sketches. His discoveries of what I meant to show in these drafts have been enlightening.

I am also pleased to acknowledge the documentary material and other help given by the manufacturers through their representatives in the U.K. The patience and forebearance of these individuals in the face of my sometimes desperate and often impossible requests over the last year or so have been more than professional etiquette demands. Accordingly I express my gratitude to: Brian McGuigan of AGA Geotronics; David Wallis of Survey and General Instrument Company Ltd.; Brian Snelling, Tony Davis and Andy Perry of Wild Heerbrugg (U.K.) Ltd.; Mike Ewer of C.Z. Scientific Instruments Ltd.; and Mark Hart of Carl Zeiss (Oberkochen) Ltd.

I am also grateful to J. McKay of A. Clarkson & Co. Ltd., N. Vyner of Keuffel & Esser Company, J. Middleton of MS Precision Ltd., R. Clarke and A. B. Evans of Hall & Watts Ltd. and G. A. Walsh of Survey Supplies Ltd., all of whom by correspondence gave me useful information in response to my requests.

I acknowledge the permission given by the following manufacturers to publish the figures listed: AGA Geotronics AB 3.42, 3.43; Carl Zeiss (Oberkochen) 3.12, 3.29, 8.2; Hall & Watts Ltd. 8.23, 8.27; Hewlett-Packard Company 3.34; Kern & Co. Ltd. 3.6, 3.11, 3.38, 3.39, 3.49, 3.51; Keuffel & Esser Company 3.44; Spectra-Physics, Inc. 8.26; VEB Carl Zeiss Jena 3.10, 3.14, 8.10, 8.21, 8.22a & b; Wild Heerbrugg Ltd. 1.8, 3.1, 3.5, 3.8, 3.9, 3.13, 3.25, 3.26, 3.31, 8.17, 8.18, 8.19, 8.20, 8.25. In addition, I am grateful to Professor D. J. Hodges of the University of Nottingham for supplying Fig. 3.35 and to the Keuffel & Esser Company for the information used to compile Fig 3.45. But, in spite of the assistance that has been given, I must claim for myself all responsibility for any errors that may appear.

Finally, but not least significantly, I wish to express my thanks and love to Jennifer, Nicholas and Jonathan. Their encouragement during the re-writing of this book has helped to do it without the realisation that I should have been engaged in less unsociable activities during my spare time.

M. A. R. Cooper

Introduction

When electronics are introduced into an older technology, one of the major difficulties, often disregarded and seldom carefully considered, is the confusing nomenclature that comes in the wake of the innovation. This confusion is not the fault of the electronics engineers, who usually have a very clear understanding of what they mean when they talk about their technology – otherwise, it would not work as well as it does. Unfortunately, practitioners of a subject into which electronics is being introduced seldom have the opportunity to develop an equivalent understanding. To begin with, they are presented with descriptions of the new technology which have been prepared with the main objective of advertising and selling a new piece of equipment. This of course is not reprehensible, but the words chosen for advertising are not always those which rigorously define a particular innovative feature. Indeed, if they were, it would probably be a poor advertisement.

The first words one reads about a new piece of equipment often stick in the memory. They are used in discussions and articles and not only in relation to the particular object to which they were first attached, but also in connection with the apparently similar but fundamentally different things. Eventually, there is a danger of complete misunderstanding as the meaning becomes more diffuse and the words degenerate to jargon.

It would be presumptuous to attempt to rectify that situation just by publishing a textbook. However, it is possible not to add to the confusion and perhaps, to clear a little of the fog of jargon that has gathered in recent years around theodolites and levels. This can be done by making here some general remarks about the meanings of certain words as they are used in this textbook.

An automatic computer which carries out a limited amount of computation and transfer of data from one device to another is referred to as a *microprocessor*. The user cannot normally program a microprocessor. It contains within its memory a set of instructions which cannot be changed by the user. It is used in theodolites and in instruments for electromagnetic distance measurement (*EDM*) to control the electronic components and to collect and correlate the signals generated by them. It is then referred to as the *controlling microprocessor*. The operator instructs the controlling microprocessor via signals generated at switches or at a keyboard, or at both.

A theodolite with optoelectronic scanning of the circles is called an *electronic theodolite*. If an EDM device is associated with an electronic theodolite and is controlled by, or connected to, the same microprocessor which controls the scanning of the circles, then the instrument is referred to as an *electronic tacheometer*.

There is a difference between automatic *registration* and automatic *recording* of data with a theodolite or tacheometer. Automatic registration is the acceptance of electrical signals from an electronic device (usually a photodiode) by the controlling microprocessor and the subsequent transformation of those signals by that microprocessor into data (binary, binary coded decimal or decimal) representing measurements. This registration might result in the temporary storage of the data in the memory of the microprocessor, or in a visible display of the decimal digits, or in both. A new pointing of the theodolite will result in those data being deleted and replaced automatically by the newly registered data. Automatic recording, on the other hand, is the storage of registered and other data on a device from which those data can later be automatically extracted – usually independently of the theodolite or tacheometer used for the measurements – and then used for automatic computation. This extraction of recorded data from the device is usually made under the control of a microprocessor or minicomputer through an interface. Automatic recording is not possible without prior automatic registration. When an electronic theodolite or tacheometer can be used to record data automatically, it is called an *electronic recording theodolite*, or *electronic recording tacheometer*.

The use of the description 'Total Station' is reserved solely for the particular instrument which has been given that name by its manufacturers. Various other electronic tacheometers and theodolites are often referred to as total stations in private discussions, in publications and at conferences, but it is hard to see in what sense any of those instruments is either total, or a station. The description 'semi-total station' (an instrument surely no manufacturer would claim to produce and no surveyor would want to buy) is thereby avoided in this book, but the author notes with dismay its growth in current usage.

Electromagnetic distance measurement is the subject of a book by C. D. Burnside in the series of which this book is a part. EDM is therefore not described here, except in so far as it is integrated either optically or electronically with a theodolite. Even then, it is described only as far as is necessary for an understanding of the way in which the angle-measuring components of the tacheometer work.

The first edition of this book was published in 1971. During the preceding few years, the introduction of electronic components into theodolites had begun with the Fennel 'code-theodolite' and the Kern 'recording tacheometer'. In each of these instruments, 35 mm film was used for registration and for recording the circle readings. The main disadvantages of film were that it had to be processed and the methods for recovering data automatically recorded on film were expensive and cumbersome. In those early days of electronic theodolites, it was decided not to include descriptions of them in the book.

Since that time, nearly all the major manufacturers of surveying equipment have developed and produced at least one electronic theodolite or tacheometer. The disadvantages of film were overcome by automatic electronic registration of data and, at first, by the use of paper tape as a medium for automatic recording of data. The Zeiss (Oberkochen) RegElta and the Aga Geodimeter 700 were produced, each having a paper tape punch for automatic data recording.

The next four electronic instruments to be produced were exhibited at the 15th Congress of the International Federation of Surveyors (FIG) held in 1977 at Stockholm. They were the Hewlett-Packard 3820A Total Station, the Kern Electronic Theodolite E1, the Keuffel & Esser Vectron and the Wild Tachymat TC1. More recently, the Kern E2, and the Zeiss (Oberkochen) Elta 4 and Elta 2 have been produced. The tendency has been towards smaller and more compact instruments with solid-state memory devices.

It is clear now that electronic components will continue to be used in theodolites and it is therefore necessary to include discussions of such components in this second edition. Several decisions had to be made about the form and contents of such discussions in relation to the rest of the book. These decisions were all made on the basis of the recognition that electronics and their introduction into a theodolite do not constitute a major change in the principles of construction of the instrument. Automatic registration of coded circle positions is but the latest step in the improvement and refinement of theodolites which have included, *inter alia*, the introduction of the telescope, light metallic alloys, glass arcs, optical micrometers and automatic optomechanical indexing of the vertical circle. It is on the basis of this view that details of the principles and application of electronic components have been included in the text.

The form of the book therefore remains unaltered. There is no additional chapter dealing with electronic instruments. Chapter 1 (Basic constructional features of the theodolite) has been enlarged to include a little historical and etymological background and a discussion of the units of angular measure. The basic components for the measurement of horizontal and vertical angles are described and their relationships defined. These features are illustrated in the case of one modern theodolite.

Chapter 2 (Principles of theodolite construction) has been greatly enlarged. The optical principles of the telescope are explained and the use of prisms, lenses and mirrors to produce a satisfactory image is described. A section on kinematic design has been added. Axis systems and movement controls are described. Methods of reading the circles are given, and these now include optoelectronic scanning of the circles as well as optical methods. It has been thought suitable to include in this section some of the elementary physics of the solid-state diode and its variations. The level of presentation is about the same as that used elsewhere in the chapter in relation to optical imaging. It is hoped that this level of presentation will satisfy the requirements of most users of theodolites. It will enable the experienced surveyor to understand in principle the ways in which electronic and other components of the theodolite work and thereby lead to a more efficient use of the newer instruments. The student should find it a useful description of the principles of design not readily found elsewhere. The specialist will see that the references quoted lead to more detailed descriptions of the principles of optoelectronic imaging, although the number of textbooks on the subject is at the moment very small. The spirit level, centring systems, methods of indexing the vertical circle, the tilt sensor and methods of referencing the vertical circle are also described. In each case, it is the principles which are emphasised in this chapter.

In chapter 3 (Features of modern theodolites) examples of specific components of modern theodolites are described to illustrate how the principles of construction described in the preceding chapter are put into effect. No particular instrument is described fully, but different components from a variety of instruments are used as illustrations. The opportunity to revise some of these illustrative examples as a result of the introduction of newer optomechanical instruments has been taken. In addition, this chapter has been greatly enlarged by the inclusion of specific examples of the applications of electronics. Circle scanning methods, a controlling microprocessor and a tilt sensor are described in detail.

Chapter 4 (Adjustments to the theodolite) contains descriptions of tests and adjustments that a theodolite user should carry out from time to time. The effects of maladjustments on the observations are described and suitable observing procedures which will reduce or eliminate these effects are given. Adjustments of the vertical axis, plate level, line of collimation, reticule, vertical circle index, telescope and the optical plummet are described. A summary is given in the form of a table.

Chapter 5 (Accuracies of constructional features of theodolites) is a discussion of the sources of error which the user cannot normally eliminate by adjustments but which can be reduced by suitable observing procedures, or by computation of 'correction' terms. The sources of error discussed are: trunnion axis dislevelment; inclinations of the circles; eccentricities of the circles; circle graduation errors; axis strain; centring errors; micrometer errors and errors in the automatic vertical circle index. The effects of these errors are summarised in the form of a table. Methods of observing and recording horizontal and vertical angles are given, with some hints to inexperienced users.

Levels are dealt with in a similar way, chapter 6 being an introduction (Basic constructional features of the level). It describes the three basic methods of achieving a horizontal line of sight: by a dumpy level; by a tilting level, and by an automatic level using a compensating device. The principles of the laser level are described.

In chapter 7 (Principles of level construction) the topics discussed are: telescopes; standing axes; movement controls; horizontal circles; spirit levels; centring; reversible levels; the quickset head; the parallel plate micrometer and automatic levels, including the principles of compensation for residual tilts.

Chapter 8 (Features of modern levels) is similar to chapter 3 in that no individual instrument is described completely. Specific components from various modern levels are described in detail to illustrate how the principles of construction are put into effect. Compensators, parallel plate micrometers, a quickset head, the dumpy level and a laser level are used as illustrations.

Tests and adjustments which the user of a level should carry out from time to time are described in chapter 9 (Adjustments and accuracies of non-automatic levels). These include adjustments to the dumpy level, the tilting level and the reversible level. The two-peg test and the three-peg test for precise levels are described, with numerical examples. Methods for reducing the effects of these errors are given. Also, suggestions for reducing the effects of errors in the staff and from the atmosphere are made. Recommendations for recording the measurements are illustrated by examples.

In chapter 10 (Errors and accuracies of automatic levels) the emphasis is on the procedures that can be adopted to reduce the effects of errors inherent in optomechanical compensators. These errors are classified and described. Their magnitudes are quoted from the results of tests carried out by several workers and published in the references. A test for determination of the errors in a laser level is described. A recent development – motorised levelling – is described and some results are given. A summary of the errors of automatic levels is given in the form of a table.

1 Basic constructional features of the theodolite

Measurement of angles is a procedure which is basic to many surveying tasks. In order to use the results of angular measurements in the computation of positions of points, the method of measurement must be related to the system which is used to define those positions. One of the most convenient ways of defining the positions of points is in terms of rectangular X, Y and Z co-ordinates, where the XY-plane is a horizontal plane and the Z-axis is vertical.

At any point on the surface of the earth, the vertical direction is the direction of the gravitational force at that point. This direction can be found by means of a plumb-bob – the plumb-bob string indicates the vertical direction. A horizontal line through a point on the surface of the earth is any line which is normal to the vertical at the point. All such horizontal lines at a point lie in a horizontal plane.

Measurement of angles between points on the earth's surface should therefore be made by reference to this horizontal and vertical co-ordinate system. Suppose A, B & C (Fig 1.1) are three points on the earth's surface and that π is the horizontal plane through A.

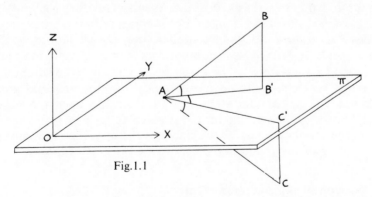

Fig.1.1

If the positions of A, B and C are of interest with respect to the axes OX, OY and OZ then the measurement of the angles at A between B and C should be designed so that the angles B'AC', BAB' and CAC' are obtained, where B′ and C′ are the orthogonal projections of B and C respectively on to the horizontal plane π. Angle B′AC′ is in the horizontal plane through A and is therefore a horizontal angle. Angles B′AB and C′AC are in planes through A normal to π (i.e. vertical planes) and these angles are vertical angles. A theodolite is an instrument designed to measure horizontal and vertical angles.

In practice, it is impossible to measure the angles at A (a point on the surface of the earth). The theodolite is placed on a tripod and is a metre or so vertically above A with the result that the horizontal angle actually measured is the same as angle B′AC′. The vertical angles measured from the tripod however are not the same as angles B′AB and C′AC respectively so an allowance has to be made in the subsequent computations for the height of the theodolite above A.

One of the earliest instruments for the measurement of horizontal and vertical angles is illustrated in the encyclopaedia *Margareta Philosophica* by Gregorius Reisch (1471–1528) and is referred to as the *Polimetrum* of Waldseemüller (1470–1518).

The first use of the word *theodolite* (as *theodolitus*) is in *A geometrical practise named Pantometria*, written by Leonard Digges (1510–50) and edited and published by his son Thomas (1546–95) in England in 1571. In *Pantometria*, the name *theodolitus* is given to an instrument designed to measure horizontal angles only, whereas an instrument for the measurement of both horizontal and vertical angles is called a *topographicall instrument*. The use of *theodolite* to describe an instrument for measurement of both horizontal and vertical angles became common shortly after the publication of Digges' work. The etymology of the word is uncertain. The most likely origin, according to the *Oxford English Dictionary* (1933), is in an unscholarly formation from the Greek $\theta\epsilon\acute{\alpha}\sigma\mu\alpha\iota$ (I view) or $\theta\epsilon\hat{\omega}$ (behold) and $\delta\hat{\eta}\lambda o\varsigma$ (visible, clear) with a meaningless termination.

The *Polimetrum* of Waldseemüller and Digges' *topographicall instrument* have certain basic components that are present in even the most modern theodolites. The materials from which the components are made have changed and technological development has greatly increased the accuracy and stability of the instrument and allowed it to be made smaller, lighter and probably easier to use. These changes are only in the details and not in the principles of construction. Some basic components of the theodolite are described in this chapter.

1.1 Horizontal angle measurement

The horizontal angle B′AC′ (Fig 1.1) is required, so a circle graduated in units of angular measure around its circumference must be placed in coincidence with the horizontal plane π. This is the *horizontal circle*, or *lower plate*. It is generally stationary during the measurement of a horizontal angle and the centre of the graduations must be placed in coincidence with the vertical through A. Setting the plane of the horizontal circle into coincidence with the horizontal plane is often called *levelling* the theodolite and placing the centre of the circle graduations vertically above the ground mark A is called *centring* the theodolite.

It is necessary to have a sighting device (a telescope) connected to an index and for these to be rotated about an axis through the centre of the circle graduations and perpendicular to the plane of the circle. The telescope and the index together form the *alidade* and their axis of rotation is the *vertical axis* (primary, main, standing or rotation axis).

In order to sight targets at different elevations, the telescope must be rotated about a second axis perpendicular to the vertical axis. This is the *trunnion* axis (secondary, transit or horizontal axis). These features are illustrated in Fig 1.2.

The *standards* support the trunnion axis and should be tall enough to enable the telescope to be *transitted*, i.e. rotated through 180° about the trunnion axis in either direction. The standards are hollow and

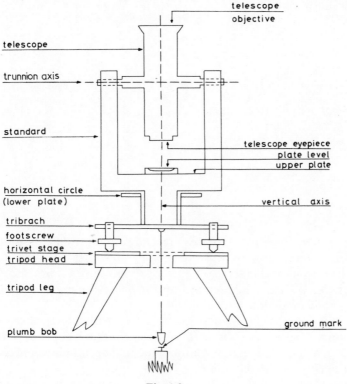

Fig. 1.2

contain optical, mechanical and sometimes electronic components for reading the circles.

To measure the horizontal angle, B'AC' (Fig 1.1) the telescope is directed towards B and the position of the index against the circle graduations noted (reading R_1). The telescope is then directed towards C and a second reading (R_2) is taken. The horizontal angle at A between B and C is then ($R_2 - R_1$).

The field of view of the telescope is a few degrees. Readings are often required to an accuracy of the order of seconds of arc, so it is not sufficient to point the telescope so that the target falls more or less in the centre of the field of view. A modification to a simple telescope has therefore to be made. A glass disc is mounted inside the telescope barrel with its plane perpendicular to the longitudinal axis of the telescope. A pattern of lines is engraved or etched on the disc (Fig 1.3). This pattern is the *reticule*. Generally, the reticule has two perpendicular lines, one in a horizontal plane and the other in a vertical plane. These two lines are often referred to as the *cross-hairs* because in older theodolites the reticule was formed by stretching hairs or threads of a spider's web across a metal diaphragm and the term *cross-hairs* has persisted.

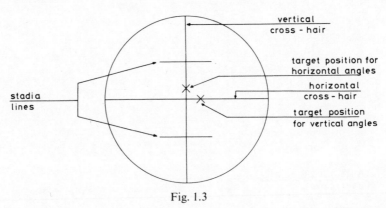

Fig. 1.3

Two shorter lines are usually present in the reticule, equidistant from the intersection of the cross-hairs and perpendicular to the vertical cross-hair. These two shorter horizontal lines are the *stadia lines* (or *stadia hairs*) and are used for the measurement of distance by vertical staff tacheometry.

The line of sight can be defined in terms of the intersection of the cross-hairs: it is the line joining this intersection to the optical centre of the objective lens. When thus defined, the line of sight is called the *line of collimation* of the telescope.

It should be noted that it is not necessary to direct the telescope for each pointing so that the target is made to appear at the intersection of the cross-hairs. Horizontal angles and vertical angles are usually observed separately, so that when the horizontal angle B'AC' (Fig 1.1) is being measured and a pointing being made to B, say, it is sufficient to direct the telescope such that B appears to lie on the vertical hair and near its centre (see Fig 1.3). The vertical hair will lie in the vertical plane BAB' and hence the correct reading will be obtained on the horizontal circle. Similarly when observing a vertical angle, the target (B, say) should appear to lie on the horizontal hair, somewhere near its centre.

1.2 Setting-up

It has been stated in section 1.1 that the theodolite must be centred and levelled. Figure 1.2 illustrates those components of the theodolite which are used to set the instrument in the correct attitude. This procedure is called *setting-up*. There are several procedures which can be used to set up theodolites. Choice of a method depends to some extent on the user's personal preference, but it is often limited by the type of theodolite and mounting being used. The division of the following setting-up procedure into separate centring and levelling stages is applicable to most theodolites, but it is not necessarily the most efficient. The manufacturer's handbook usually describes a method which is convenient to use with a particular instrument.

1.2.1 Centring

The requirement is that the centre of the circle graduations lies vertically above the ground mark at which observations are needed.

A plumb-bob is suspended from a point on the vertical axis of the theodolite, the point of suspension indicating the centre of the circle graduations. The *trivet stage* (or stage plate) carrying the theodolite is moved laterally on the *tripod head* until the plumb-bob lies over the ground mark. The trivet stage is then clamped to the tripod head. It is important to set the tripod so that it is approximately centred before attaching the theodolite to it.

1.2.2 Levelling

The requirement is that the axis of rotation of the alidade (the vertical axis) is vertical. This axis is normal to the plane of the horizontal circle, so levelling ensures that this plane is horizontal.

The vertical axis can be tilted in any direction by means of the *footscrews*. These stand on the trivet stage and by turning them, rotation is transmitted via the *tribrach* to the vertical axis. A spirit level (*the plate level*) attached to the alidade serves to indicate when the vertical axis is vertical. The horizontal circle index, the plate level and their connecting parts are often called the *upper plate*.

1.3   Vertical angle measurement

Vertical angles such as B′AB and C′AC (Fig 1.1) are required, so a circle graduated in the appropriate units of angular measure around its circumference must be placed in a vertical plane. This is the *vertical circle*.

The vertical circle has its centre of graduation on the trunnion axis, and it rotates with the telescope about this axis. The vertical circle index remains stationary. Vertical angles (such as B′AB and C′AC in Fig 1.1) are all measured from a horizontal plane, so a representation of this plane must be present in the vertical angle reading system. The line joining the vertical circle index to the centre of the vertical circle graduations gives this representation when it is horizontal. In order to make it horizontal, the *altitude level setting screw* is used. When this is adjusted, only the frame (shaded in Fig 1.4) rotates about the trunnion axis.

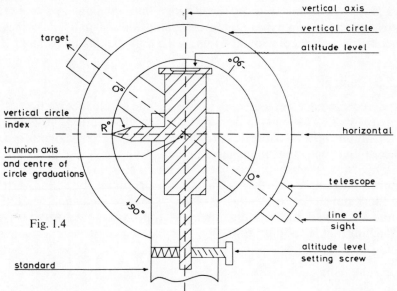

Fig. 1.4

This frame carries the index and the *altitude level*. When the bubble of the altitude level is central, the line joining the index to the centre of the vertical circle graduations is horizontal so the circle reading (R°) is the vertical angle.

The position of the vertical circle when viewed from the eyepiece end of the telescope is used to describe the attitude of the theodolite; if it is on the left of the telescope (Fig 1.5a) the theodolite is said to be in the *face-left* (*F.L.*) attitude and if it is on the right of the telescope (Fig 1.5b) the theodolite is in the *face-right* (*F.R.*) attitude.

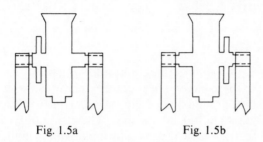

Fig. 1.5a   Fig. 1.5b

To change face, the telescope is rotated 180° about the trunnion axis (i.e. it is transitted) and then the alidade is rotated 180° about the vertical axis. This operation is nearly always used because it eliminates the effects of some instrumental errors. The alidade should always be rotated by the standards and not by the telescope, to reduce strain on the trunnion axis.

It has already been mentioned in the introductory paragraphs to this chapter that the vertical angle measured from a tripod above a ground mark is not the same as that required, which is the angle B′AB say, in Fig 1.1 and that an allowance for the height of the instrument has to be made in subsequent computations involving the measured vertical angle. It can now be seen that the vertical angle measured is the angle relative to the horizontal plane through the centre of the vertical circle graduations. Thus the height required to correct for this in computations is the vertical distance from the ground mark A to the centre of these graduations, that is to the trunnion axis.

Unlike the horizontal circle which is usually graduated clockwise around its circumference from 0° to 360° (or 0^g to 400^g) the vertical circle can be graduated in a number of different ways, some of which are illustrated in Figs 1.6a, b and c.

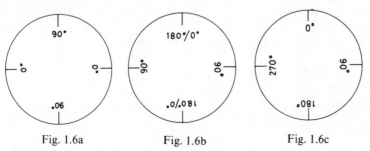

Fig. 1.6a   Fig. 1.6b   Fig. 1.6c

If an angle of depression of 1° is observed, circle readings would be as follows:

Type	Face Left	Face Right
1.6a	1°	1°
1.6b	91°	89°
1.6c	91°	269°

1.4 Theodolite axes

There are three basic axes in a theodolite and from the foregoing sections it can be seen that these have to bear certain relationships to each other. These relationships are illustrated diagrammatically in Fig 1.7.

The vertical axis should be vertical. The trunnion axis should be perpendicular to the vertical axis and hence horizontal. The line of sight should be perpendicular to the trunnion axis and should pass through the intersection of the vertical axis and the trunnion axis.

1.5 Movement controls

The following two rotations are always possible:

(a)   Rotation of the alidade about the vertical axis. This also rotates the trunnion axis and the line of sight about the vertical axis.

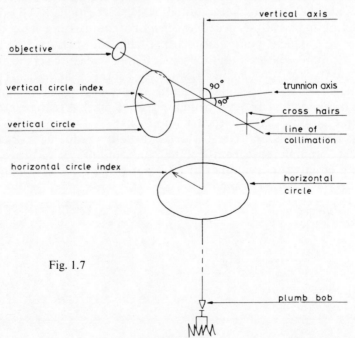

Fig. 1.7

(b)   Rotation of the telescope (and hence the line of sight and the vertical circle) about the trunnion axis.

In addition, there is generally some provision for the rotation of the horizontal circle about the vertical axis. Generally each of these rotations can be carried out by a *slow motion screw* (or tangent screw) in either direction through a range of a few degrees. If a large rotation has to be made, then a clamp can be released and the particular component can be rotated freely about its axis. The clamp has to be on before the slow motion screw drive is engaged. The slow motion screws enable accurate setting of the alidade or the telescope to be made.

1.6 Circle graduations and reading systems

There are three common units of angular measurement. The fundamental unit is the *radian* which is the angle subtended at the centre of a circle by an arc equal in length to the radius of the circle. There are therefore 2π radians in a full circle – the angle subtended at the centre by the circumference. The fundamental importance of the radian is however not sufficient to make it also a unit of practical importance because significant angles, such as a right-angle, are expressed as irrational numbers of radians.

The *sexagesimal system* was in use in early Babylonian times. It has been used continuously in astronomy and navigation since then and by cartographers since at least the time of Ptolemy. It is the basis of the graduations on the *Polimetrum* of Waldseemüller. There are three hundred and sixty *degrees* (360°) in a circle, so a right-angle contains 90°. A degree is divided into sixty *minutes* (1° = 60′). A minute is divided into sixty *seconds* (1′ = 60″). The sub-division of a second is by decimal fractions. There are 3600″ in 1°. An angle in sexagesimal measure is written in the form 67° 42′ 38.6″, for example.

In France, during the first half of the nineteenth century, rationalisation of standards of measurement was introduced, although general acceptance of the new units was slow and reluctant. At this time, the *centesimal* system was codified and is now in common use in several European countries and elsewhere. There are four hundred *grades*, or grads(400^g) in a circle, so a right-angle contains 100^g. A grade is divided into one hundred *centigrades* ($1^g = 100^c$). A centigrade is divided into one hundred centi-centigrades ($1^c = 100^{cc}$). Further sub-division is by decimal fractions. A centigrade and a centi-centigrade are sometimes referred to as a centesimal minute and a centesimal second respectively. An angle in centesimal measure is not usually written as $384^g\,26^c\,67^{cc}$, for example, but as the decimal fraction 384.2667^g. Subdivision of the basic unit by factors of 10^2 is not in accordance with the Système Internationale (SI) which specifies that subdivision of the basic unit shall be by factors of 10^3. For this reason, the *milligrade* (or milligrad) has come into use (1 grad = 1 000 mgrad). In modern usage, the grade is referred to as the *gon*. Subdivision by 10^3 gives the *milligon* (1 gon = 1 000 mgon).

Centesimal values and useful approximations of the sexagesimal units are as follows:

$$1° = 1.111\ 111.....................^g \fallingdotseq 1.1^g \text{ or } 1.1 \text{ gon}$$
$$1' = 1.851\ 851.....................^c \fallingdotseq 1.85^c \text{ or } 18.5 \text{ mgon}$$
$$1'' = 3.086\ 419\ 753\ 086.......^{cc} \fallingdotseq 3.1^{cc} \text{ or } 0.31 \text{ mgon}$$

Sexagesimal values (exact) of the centesimal units are as follows:

$$1^g = 0.9° \quad = 54'$$
$$1^c = 0.54' \quad = 32.4''$$
$$1^{cc} = 0.324''$$
$$1 \text{ mgon} = 3.24''$$

The number of seconds in a radian (usually denoted by ρ'' or sometimes written as cosec $1''$) is:

$$360 \times 60 \times 60 \times 2\pi \fallingdotseq 206264.8$$

to seven significant figures. This value and similar values for the other sexagesimal and centesimal units are useful for the conversion of an angle expressed in radians to the equivalent in practical units. Approximate values (accurate to seven significant figures) are:

$$\rho'' = 206\ 264.8 \fallingdotseq 2.1 \times 10^5$$
$$\rho' = 3\ 437.747 \fallingdotseq 3.4 \times 10^3$$
$$\rho° = 57.295\ 78 \fallingdotseq 57$$
$$\rho^{cc} = 636\ 619.8 \fallingdotseq 6.4 \times 10^5$$
$$\rho^c = 6\ 366.198 \fallingdotseq 6.4 \times 10^3$$
$$\rho^{gon} = 63.661\ 98 \fallingdotseq 64$$
$$\rho^{mgon} = 63\ 661.98 \fallingdotseq 6.4 \times 10^4$$

For any angle ϕ, $\phi° = \phi\text{rad} \times \rho°$
and $\phi'' = \phi\text{rad} \times \rho''$ etc.

A fourth, but less common, unit of angular measurement is the *millième*, often abbreviated to *mil*. There are 2π radians or approximately 6 283 *milliradians* (mrad) in a circle. A millième is defined as 1/6 400, or sometimes as 1/6 000, of a complete circle, so it is approximately equal to 1 milliradian. The main reason for using this unit is to take advantage of the simple relationship between arc length, radius and subtended angle which is inherent in the definition of a radian: 1 mil subtends a length of approximately 1 unit at a distance of 1 000 units. This simple, approximate relationship has been used for military purposes, mainly in connection with artillery, but use of the millième is obsolescent and it is rarely used as a unit for theodolite circle graduations.

There are three main methods of reading a theodolite circle: mechanical, optomechanical and optoelectronic. In older theodolites, the circle graduations are etched or engraved on metal and exposed, at least over a small arc of circumference, so that they can be read by eye using a vernier or a micrometer, sometimes with the aid of a simple magnifier or small microscope. More recently, since about 1920, the graduations have been made by the photochemical, or vacuum, deposition of thin metallic film on glass arcs and reading is by eye using an optical scale or optical micrometer. Illumination of the glass arcs and reading scales is either by small adjustable plane mirrors attached to the exterior casing of the theodolite to reflect natural light, or by small electric light bulbs. Most recently, reading is by optoelectronic scanning of the circles. In this case, the circle readings are presented to the operator by electronic displays after conversion of the analogue signal from the scanning device to digital equivalents.

1.7 Actual examples

Figure 1.8 illustrates some of the components described in this chapter. The theodolite illustrated is the Wild T2.

Fig. 1.8

Fig 1.8 Key

1. eyepiece for optical plummet
2. tribrach
3. illumination mirror for horizontal circle
4. support point for carrying in container
5. horizontal clamp for alidade
6. vertical slow-motion screw
7. optical sight with point for centring under roof-points
8. vertical clamp
9. illumination mirror for vertical circle
10. telescope objective
11. safety catch for carrying handle
12. carrying handle
13. locking screw for carrying handle
14. lever for reticule illumination using lighting kit
15. optical micrometer knob
16. telescope focusing sleeve
17. eyepiece diopter scale
18. eyepiece of circle reading microscope
19. eyepiece of telescope
20. selector knob for viewing either horizontal or vertical circle
21. plate level
22. horizontal slow motion screw for alidade
23. cover for horizontal circle drive knob
24. spherical level
25. clamp connecting measuring parts to the tribrach
26. footscrew
27. trivet stage
28. spring plate

2 Principles of theodolite construction

The basic constructional features of the theodolite which were outlined in the previous chapter are now considered in more detail.

A theodolite is an arrangement of mechanical, optical and electronic components which must be assembled accurately. This is a requirement which has to be met by many other instruments used for measurement. However, theodolites are used in conditions in which few other similarly precise instruments are required to be used. Any theodolite must give accurate results (often to the order of 1″) in both tropical and arctic weather. It is often transported in vehicles over rough country and must stand up to the knocks, dust and damp which such travelling often brings. It must be small and light enough to be carried comfortably, often for several miles and yet strong enough for the carrier not to be required to take an undue amount of care. A theodolite is often used on construction sites or underground, close to heavy vibrating machinery and must be capable of high-precision measurements under such conditions. It is often used in areas far from suitable servicing and repair facilities and must therefore be capable of maintaining its original precision with only minor adjustments for long periods of use. Despite these demands, the major manufacturers have succeeded, often in different ways, in manufacturing theodolites which meet these requirements of the user.

In this chapter, only the principles of theodolite construction are considered, under the headings telescope; vertical axis systems and kinematic design; controls for movement in azimuth; circle reading systems; spirit levels; centring systems; and vertical circle indexing systems. Examples of how different manufacturers have put these principles into effect are described in the following chapter.

2.1 The telescope

2.1.1 Image formation

Figure 2.1 illustrates the basic features of a surveyor's telescope but

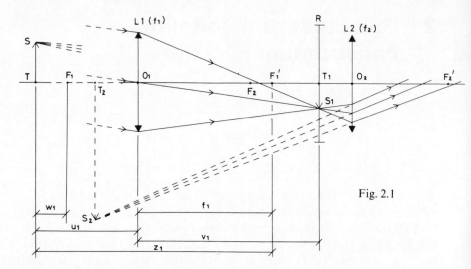

Fig. 2.1

it will be shown that these features are modified considerably in practice. The lenses are assumed to be thin.

Two converging lenses L1 (the objective) and L2 (the eyepiece) with focal lengths f_1 and f_2 respectively are arranged so that their axes passing through the optical centres 0_1 and 0_2 respectively are coincident. The objective has its primary focal point at F_1 and its secondary focal point at F_1'. The primary and secondary focal points of L2 are at F_2 and F_2' respectively.

A small object ST perpendicular to the common axis $0_1 0_2$ has a real inverted image $S_1 T_1$ (the intermediate image) formed by the objective. The reticule R, carrying the cross-hairs is situated in the plane of the intermediate image $S_1 T_1$; the objective is moved along the axis $0_1 0_2$ so that images of objects at different distances are always formed at the reticule.

The positions of the object ST and the intermediate image $S_1 T_1$ are interdependent and are given in Gaussian form by the equation

$$\frac{1}{f_1} = \frac{1}{u_1} + \frac{1}{v_1}$$

and in Newtonian form by

$$w_1 z_1 = f_1^2$$

where f_1 = focal length of objective $w_1 = F_1 T$
       $u_1 = 0_1 T$ (object distance) $z_1 = F_1' T$
and $v_1 = 0_1 T_1$ (image distance).

For nearly all the object distances encountered in surveying, the intermediate image will be formed very close to the secondary focal point F_1' of the objective and in all cases the objective must be moved so that the intermediate image is formed in the plane of the reticule. However, it is shown in section 2.1.7 that this arrangement has several disadvantages, and in modern telescopes a third (internal) lens is used for focusing.

For an objective L1 of focal length 200 mm, Fig 2.2a shows the relative positions of the objective and reticule R with an object at infinity. As the object is moved closer to the theodolite, the separation between the objective and the reticule must be increased by moving either the objective away from the reticule or the reticule away from the objective. When the object is 5 m from the objective, Fig 2.2b illustrates the relative positions of objective and reticule.

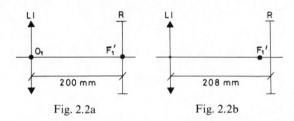

Fig. 2.2a   Fig. 2.2b

The cross-hairs and the intermediate image S_1T_1 (Fig 2.1) serve as an object for the eyepiece L2. The position and nature of the final image S_2T_2 will depend upon the position of the eyepiece. This should be set so that the final image is between the *near point* (or least distance of distinct vision) and the *far point* (or greatest distance of distinct vision). These points vary from one individual to another and for one individual they vary with age. For a normal observer, the former is about 200 mm from the eye and the latter at infinity. It is better to adjust the eyepiece so that the final image is at the far-point since the ciliary muscles will then be most relaxed and the observer will be comfortable for a longer time. This adjustment is described in section 4.5.1.

Equations analogous to those above give the relation between the positions of the intermediate and final images;

$$\frac{1}{f_2} = \frac{1}{u_2} + \frac{1}{v_2} \text{ and } w_2z_2 = f_2^2$$

Either of these two equations shows that an eyepiece of 10 mm focal length must be positioned approximately 9.6 mm from the intermediate image if the final image is to be at a near-point of 250 mm, or of course 10 mm away if the final image is to be at infinity.

It can be seen that for the telescope described, the overall length will have to be at least 218 mm, which is rather long. Shorter telescopes can be made if the focusing is carried out by an internal diverging lens as described in section 2.1.7.

2.1.2 Magnification and Aperture

The size of a retinal image is proportional to the angle subtended by the object at the eye. The angular magnification of a telescope is then defined as the ratio of the angle subtended at the eye by the final image to that subtended by the object. The distant object ST in Fig 2.3 has a final image at infinity.

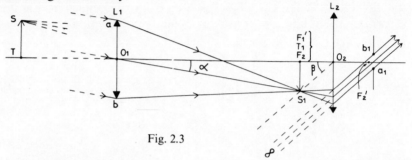

Fig. 2.3

The intermediate image S_1T_1 is approximately at F_1' (ST is a distant object) and is also at F_2.

The angular magnification M is given by

$$M = \frac{\beta}{\alpha}$$

but

$$\tan \alpha = \frac{S_1T_1}{0_1T_1} \text{ and } \tan \beta = \frac{S_1T_1}{0_2T_1}$$

therefore,

$$M = \frac{\beta}{\alpha} \simeq \frac{\tan \beta}{\tan \alpha} = \frac{0_1T_1}{0_2T_1} = \frac{f_1}{f_2}.$$

Thus

$$M \simeq \frac{f_1}{f_2}$$

The conical bundle of light rays emanating from point S on the distant object falls on the objective L1 which limits the bundle to a diameter ab. The objective therefore acts as an aperture stop and is the entrance pupil. The exit pupil is a circle (the Ramsden circle) of diameter a_1b_1 which is the image of the aperture stop formed by the eyepiece. If D is the diameter of the entrance pupil (the objective) and d that of the exit pupil then

$$\frac{f_1}{f_2} = \frac{D}{d}, \text{ which is the linear magnification}$$

(This can be seen if the marginal ray $aF_1'a_1$ is traced on the diagram.) Hence an alternative expression for the angular magnification is

$$M \simeq \frac{D}{d}$$

The magnification of any telescope can be found by measuring the two diameters. The diameter of the exit pupil can be found by pointing the telescope to the sky and setting the eyepiece to infinity. The Ramsden circle can be found and measured on a sheet of graph paper held in the correct position behind the eyepiece.

It appears then that the magnification of a telescope with a given objective can be increased to any desired value by decreasing the diameter of the Ramsden circle, by decreasing f_2. However, other factors affect the suitability of an image for surveying purposes and these are dealt with in the next sections.

2.1.3 Image Resolution

From a consideration of diffraction patterns produced by a circular aperture, the minimum angle of resolution in seconds is given by $\theta'' = (141/D)$ is the diameter of the objective in mm. For a telescope with a 40 mm objective aperture, the resolution is about 3 1/2″. The human eye has a pupil diameter of about 2.5–3 mm so the resolution of the eye is (according to the equation) about 47″. However, in practice, defects in the retina limit the resolution to about 1′. With an objective giving a resolution of about 3″, an eyepiece giving an angular magnification of $\times 20$ will enable the eye fully to appreciate the resolution. A greater magnification will be wasted, resulting in a blurring of the image and pointing errors. A smaller magnification will not take full advantage of the resolution given by the objective.

2.1.4 Brightness and Contrast

Assuming no losses in the optical system, the brightness of an image is the same as that of the object (irrespective of magnification) if both are in the same medium and if the pupil of the eye is the aperture stop. The reader should not confuse direct viewing of an image with viewing an image on a screen; in the latter case, the brightness of the image always increases as the magnification decreases. For direct viewing, if the Ramsden circle is smaller than the diameter of the eye pupil, the brightness of the image will be less than that of the object. Increasing the Ramsden circle results in a brighter image up to the point where the diameter of the Ramsden circle is equal to that of the eye pupil. A further increase in the diameter of the Ramsden circle will not result in the perception of greater brightness by the eye.

Magnification is given by $M \simeq (D/d)$ (section 2.1.2) so for a given D, the magnification can be increased by decreasing d until it is equal to the diameter of a normal eye pupil (about 2.5 mm). In practice, there will be losses in the optical system and the image brightness will not be the same as that of the object. These losses arise mainly through unwanted reflections at the surfaces of lenses. Moreover, when the stray light from these reflections reaches the image, contrast is reduced. Light reflected from a refracting surface can be reduced to about 1/10th of its normal value by coating the surface with a thin film of metallic fluoride or silica. The thin film causes destructive interference between the reflected rays which results in an improved contrast in the image. The thickness of the coating is usually designed so that light of wavelengths near the centre of the visible spectrum is reduced considerably but red and violet light is still reflected. This is why coated lenses generally appear purple in colour.

2.1.5 Field of view

This is the angle subtended at the objective by that portion of the horizon which can be seen through the telescope. It can be found by taking horizontal circle readings to an object firstly at the extreme left-hand side of the field of view and then at the right-hand side. The difference in circle readings is the actual field of view(α). The apparent field of view (α') is the angle subtended at the eye by the image of the actual field of view. Thus magnification

$$M \simeq \frac{\alpha'}{\alpha}.$$

For a given value of α' an increase in the magnification results in a decrease in the actual field of view. The field of view depends upon the size of the objective.

The following table shows the variation in magnification with different eyepieces for a telescope with an objective of 60 mm aperture and 350 mm focal length.

Eyepiece focal length (f_2)	Ramsden circle (d)	Magnification (M)
8.6 mm	1.5 mm	× 40
11.7 mm	2.0 mm	× 30
14.6 mm	2.5 mm	× 24

2.1.6 Summary of image characteristics

The table below shows how the magnification, resolution and image brightness are interdependent and how each can be improved.

A balance has to be struck between magnification, resolution and brightness. A larger objective will always improve the image provided it is matched by the eyepiece, but is of course more expensive. In general, for a given objective aperture D, it is inadvisable to have a magnification greater than 2/3 D (D in mm) since this results in a Ramsden circle of 1.5 mm diameter, slightly smaller than the diameter of the eye pupil. A higher magnification will give less brightness and a loss of resolution by the eye. In poor observing light a magnification of 1/2 D would result in a brighter image but it is only the better quality theodolites which have alternative eyepieces.

To increase	Increase the	Decrease the	which also causes
Magnification	Objective focal length		Higher resolution Increased brightness
		Ramsden circle (decrease f_2)	Loss of brightness if $d < 2.5$ mm approx.
Resolution	Objective aperture		Increased brightness Greater magnification
Image brightness	Objective aperture		Greater magnification Higher resolution
	Ramsden circle (up to 2.5 mm) i.e. increase f_2		Lower magnification

2.1.7 The internally-focusing telescope

The telescope described in section 2.1.1 is externally-focusing because either the objective or the reticule and eyepiece must be moved, usually by rack and pinion, so that the images of objects at different distances from the objective always fall in the plane of the reticule. Such a telescope is simple, but has several disadvantages;

(a) it cannot be sealed against moisture and dust,
(b) its balance changes as the position of the objective changes,
(c) the length of the telescope changes,

    (d)   coincidence between the optical axes of the lenses is likely to be disturbed, and collimation is likely to change with object distance,

    (e)   for a satisfactory magnification, the telescope will be long, necessitating tall standards with increased weight and instability

and

    (f)   its use as an instrument for stadia tacheometry involves an additive constant.

All these disadvantages are overcome by using a movable internal diverging lens to place the intermediate image in the plane of the reticule. This image is then examined by the eyepiece in the same way as in the externally-focusing telescope.

In figure 2.4 a distant object ST has an image $S_1'T_1'$ formed by the objective L1. The focusing lens L3 however, displaces this image towards the right and by suitable positioning of L3 the intermediate image S_1T_1 is made to fall on the reticule R. This image is then magnified by the eyepiece L2.

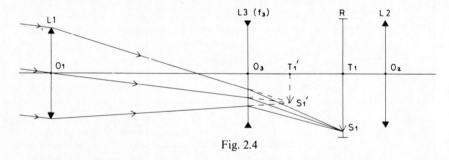

Fig. 2.4

To appreciate how such a system leads to a shorter telescope, consider an object at infinity (Fig 2.5).

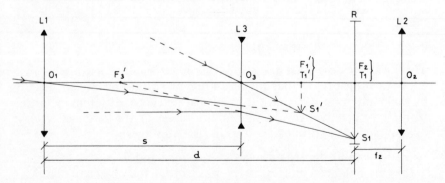

Fig. 2.5

Suppose (as in the Wild T2) $f_1 = + 118.6$ mm
$$f_3 = - 43.9 \text{ mm}$$
$$s = 0_10_3 = 98.3 \text{ mm (for } \infty \text{ focus)}$$
$$d = 136.0 \text{ mm}$$

The resultant focal length of L1 and L3 is given by

$$\frac{1}{F} = \frac{1}{f_1} + \frac{1}{f_3} - \frac{s}{f_1 f_3}. \quad \text{Thus } F = + 220.6 \text{ mm}$$

The focal length of the eyepiece (f_2) is $+ 7.9$ mm so that the magnification is

$$M = \frac{F}{f_2} = \frac{220.6}{7.9} = \times 28$$

Therefore the length from objective to eyepiece is $(d + f_2)$ i.e. $136.0 + 7.9 = 143.9$ mm.

If the telescope were externally-focusing with the same magnification, the length would be $220.6 + 7.9 = 228.5$ mm. There is thus a reduction in length of about $1/3$ for the same magnification.

The following table shows the displacement Δ_i of the internally-focusing lens L3 to the right of its infinity focus position for various object distances, u. The table is based on figures given above.

u(m)	(0.119)	1.5	2	5	10	20	50	100	200	300	500	∞
Δ_i(mm)	∞	14.3	10.5	4.06	2.00	1.00	0.40	0.19	0.10	0.07	0.04	0

These figures can be compared with the displacement Δ_e (to the left) of the externally-focusing objective of the same focal length as the above combination.

u(m)	1.5	2	5	10	20	50	100	200	300	500	∞
Δ_e (mm)	38.0	27.4	10.2	4.98	2.46	0.98	0.39	0.24	0.16	0.10	0

Thus not only is the internally-focusing lens displacement less than half that of the externally-focusing objective for a given object distance, but the movement takes place inside the telescope barrel. The coincidence between the optical axes of the lenses is therefore much more accurately maintained in the former system.

Furthermore, if the focusing lens does move perpendicular to its axis by a small amount x, the intermediate image will be displaced in that direction by less than x in the case of an internally-focusing telescope but by x in the case of an externally-focusing telescope.

2.1.8 Image defects and their reduction

So far, the telescope has been described in terms of single thin lenses. If this were the case, the final image would be of no practical use. The image is subject to five monochromatic aberrations (spherical aberration, coma, astigmatism, curvature and distortion) and chromatic aberration. Of these, the first two and the last one are most important in surveying telescopes since they affect the image near the centre of the field of view. Defects of the image in the region outside the stadia lines are not as important.

Sections 2.1.8.1–6 inclusive contain only brief descriptions of the aberrations. A fuller introduction to the subject is given by Levi, 1968.

2.1.8.1 Spherical aberration

Referring to Fig 2.6, an incident parallel light beam meeting the convex lens L will not be brought to a point focus. In effect, the focal length varies with distance from the principal axis AA′ and lies between OF_m (the focal length for marginal rays) and OF_p (the focal length for paraxial rays). $F_m F_p$ is a measure of the longitudinal spherical aberration which can be as much as a few centimetres. To reduce spherical aberration, a lens is 'bent' by having its radii of curvature altered to give an optimum condition. Alternatively a doublet can be used such that the spherical aberrations of the component converging and diverging lenses tend to cancel.

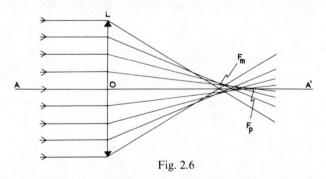

Fig. 2.6

Figure 2.7 shows the relation between focal length and radial distance for an objective doublet composed of two lenses such as those in Figure 2.8. In surveying telescopes, bisection of the target is always

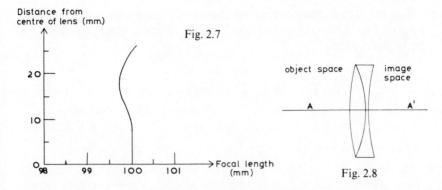

Fig. 2.7

Fig. 2.8

made near the optical axis so the doublet can be considered to be free of spherical aberration. It is then referred to as an *aplanatic doublet.*

2.1.8.2 Coma

This is the aberration of an image which lies off the lens axis. Figure 2.9 shows the nature of the aberration. In effect, this gives rise to different magnifications for different zones of the lens. When stadia readings are being made, the presence of coma causes blurring of the staff image and it therefore has to be reduced. Shaping a lens to eliminate coma will often appreciably reduce spherical aberration as well. Hence it is possible to design a doublet that will reduce both to acceptable limits.

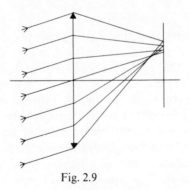

Fig. 2.9

2.1.8.3 Astigmatism

For a given object point, the ray passing through the centre of the lens is the *principal ray.* The plane containing this ray and the axis of the lens is the *meridional plane.* So far in this discussion of lens aberrations, the defects have been described in terms of rays in the meridional plane only. It is necessary also to consider aberrations arising from rays that do not lie in the meridional plane. The plane that is normal to the meridional plane and also contains the principal ray is the *sagittal plane*.

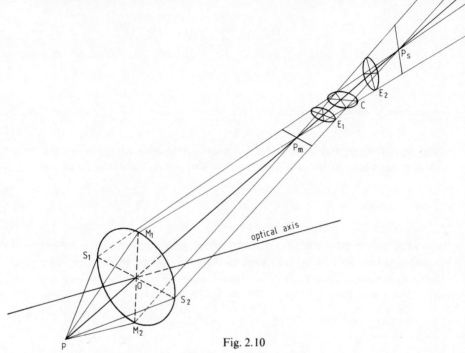

Fig. 2.10

In Fig 2.10, $POP_m P_s$ is the principal ray for the lens $S_1 M_1 S_2 M_2$ and the object point P. The meridional plane is PM_1OM_2 and the meridional rays PM_1 and PM_2 converge at P_m. The sagittal plane is PS_1OS_2 and the sagittal rays PS_1 and PS_2 converge at P_s. Because of the different focal points for the meridional and sagittal rays, the image of P is said to suffer from astigmatism and will not be a point. If a screen is placed normal to the optical axis and passing through P_m, the image of P will be a straight line tangential to a circle in the plane of the screen and with its centre on the optical axis. As the screen is moved away from the lens, the image of P will become elliptical (at E_1 for example) then circular at C, then elliptical (at E_2 for example) then a straight line at P_s, radial to the imaginary circle on the screen.

If a wheel were placed with its centre on and its plane normal to the optical axis of the lens, astigmatism would cause the image of the wheel in a plane through P_m normal to the optical axis to be sharp around the rim but blurred along the spokes. Conversely, at a similar plane through P_s, the image would be sharp along the spokes and blurred around the rim.

Astigmatism in an aplanatic doublet can be reduced either by using a stop or a third lens. In surveying telescopes, a stop would reduce the image brightness significantly, so the objective usually has three components, free of significant spherical aberration, coma and astigmatism. Such a component is an *anastigmatic triplet*.

2.1.8.4 Curvature of field

A triplet designed to reduce spherical aberration, coma and astigmatism will form point images of points on the surface of a plane normal to the optical axis. However, these point images will lie on a curved surface (the Petzval surface) and not on a plane. This aberration is effectively controlled by making the Petzval surface approximately flat in the design of the anastigmat.

2.1.8.5 Distortion

This results from a variation in lateral magnification with distance from the optical axis. If the magnification decreases, the image of a square grid will be as shown in Fig 2.11a and if it increases, it will be as shown in Fig 2.11b. The former results in barrel distortion and the latter in pincushion distortion.

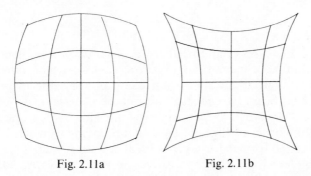

Fig. 2.11a   Fig. 2.11b

It becomes a problem when stops are introduced into lens systems, but in surveying telescopes it is of little consequence because the other aberrations are generally not corrected by stops and distortion effects on the perimeter of the field of view do not interfere with observations.

2.1.8.6 Chromatic aberration

The refractive index of a medium is a function of the wavelength of the light passing through it. Therefore the focal length and magnification of a lens vary with the colour of light passing through it. Figure 2.12a shows how parallel incident rays of white light are refracted through a converging lens to give a series of images corresponding to different wavelengths.

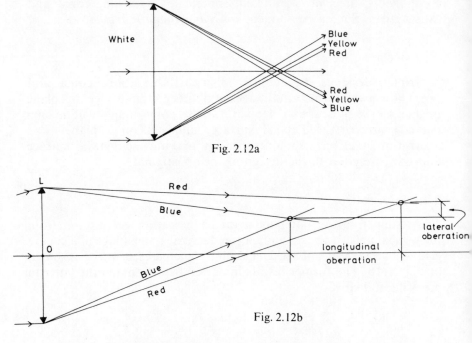

Fig. 2.12a

Fig. 2.12b

Figure 2.12b shows that this aberration has two components; lateral and longitudinal chromatic aberration, each of which must be corrected, the longitudinal aberration being of the same magnitude as spherical aberration. In general, chromatic aberration is reduced to insignificant proportions by using two lenses in contact, one of crown glass and the other of flint glass, the latter having a greater dispersive power than the former for a given focal length. If then a crown glass converging lens with a large (positive) power is used with a flint glass diverging lens with smaller (negative) power, the combination will have a positive power and effectively zero dispersion. Such a combination is an achromatic doublet and at the same time, the radii of curvature of the lenses can be selected so that the combination is approximately aplanatic. Such a compound lens could be used as an objective.

Another method of obtaining an achromatic system is to use two lenses of the same glass separated by half the sum of their focal lengths. Such a system is known as a separated doublet and could be used as an eyepiece (the Ramsden eyepiece). Each lens has the same focal length (f) so the separation for achromatism should be f (Fig 2.13).

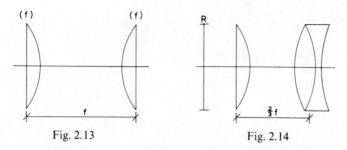

Fig. 2.13   Fig. 2.14

This is unsatisfactory because the focal plane of the doublet is at the field lens. Therefore any specks of dirt on the lens would appear in sharp focus. In order to move the focal plane outside the combination, the separation of the lenses is decreased to 2/3 f. It is thus not achromatic but its defects can be corrected to some extent by making the eye lens a doublet. This is the Kellner eyepiece (Fig 2.14). The reticule is outside the combination, so the cross-hairs will also be corrected for chromatic aberration.

A more detailed discussion of achromatism in components of surveying instruments is given by Smith, 1970, in a companion volume.

2.1.9 Anallactic telescopes

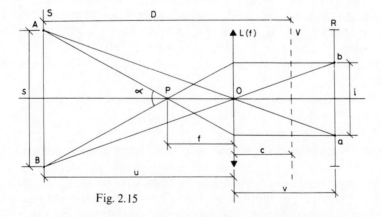

Fig. 2.15

Figure 2.15 shows a simple externally-focusing telescope with objective L of focal length f. R is the reticule carrying stadia lines a and b separated by a distance i. V is the vertical axis, distance c from the objective. A vertical staff S, at a distance D from the vertical axis is viewed through the telescope. The portion of this staff intercepted between the stadia lines is s = AB, the staff intercept.

The Gaussian lens equation is $\dfrac{1}{f} = \dfrac{1}{u} + \dfrac{1}{v}$ and $D = c + u$.

These two equations, together with the relation

$$\frac{u}{v} = \frac{s}{i} \text{ give } D = (c + f) + \frac{fs}{i}.$$

The ratio $\dfrac{f}{i}$ $\left(= \dfrac{PO}{ab}\right)$ is constant (the multiplying constant).

The term $(c + f)$ is almost constant (the additive constant) c varying as the objective is moved to focus objects at different distances onto the reticule (see Figs 2.2a and 2.2b).

$$\text{Putting } \frac{f}{i} = C \text{ and } (c + f) = k,$$
$$D = Cs + k$$

In Fig 2.15, P is the parallactic point and α is the parallactic angle. If P can be made to coincide with the intersection of the vertical axis V and the optical axis for all object distances u, the telescope is said to be anallactic and P becomes the anallactic point. The word anallactic means (approximately) 'no change' and comes from $\dot{\alpha}\nu-\dot{\alpha}\lambda\lambda\alpha\kappa\tau\kappa\varsigma$ (Lenox-Conyngham, 1942). In other words, there is no change in the position of the parallactic point.

Ignazio Porro (c. 1830) constructed the first anallactic telescope. An additional convex lens (the anallactic lens) of focal length f_1 was introduced into the telescope at a fixed distance t from the objective. Focusing was by means of the eyepiece and reticule.

It can be shown (Allan, Hollwey and Maynes, 1968, for example) that

$$D = \left[\frac{ff_1}{(f_1 + f - t)i}\right] s$$

There is no additive constant and f, f_1, i and t are chosen so that $D = 100s$.

Modern internally-focusing telescopes can be designed to have nearly anallactic properties, so Porro's modified externally-focusing telescope is no longer necessary. Considerations leading to an appreciation of anallactic design in modern telescopes follow and are based on Ollivier, 1963.

In Fig 2.16, L1 is the objective, focal length f_1, centre 0_1,
 L2 is the focusing lens, focal length $-f_2$, centre 0_2,
 V is the vertical axis,
 R is the reticule with stadia lines at a and b, separation i,

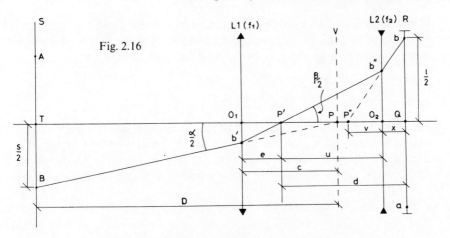

Fig. 2.16

ST is the vertical staff, intercept AB (= s),
P is the anallactic point, distance c from the objective,
α is the anallactic angle,
P″ is the image of P′ in L2, and
P′ is the image of P in L1.

For the lens L2, $\dfrac{1}{u} - \dfrac{1}{v} = \dfrac{1}{f_2}$ so $v = \dfrac{uf_2}{u + f_2}$. $\qquad$ (1)

By similar triangles, $\dfrac{\frac{i}{2}}{O_2b''} = \dfrac{v + x}{v} = \dfrac{1 + x}{v}$. $\qquad$ (2)

But $O_2b'' = u \tan \dfrac{\beta}{2} \simeq \dfrac{u\beta}{2}$ (to 2nd order).

Substitution for O_2b'' and v in (2) gives $i \simeq u\beta \left(1 + \dfrac{xu + xf_2}{uf_2}\right)$

$$\simeq \dfrac{\beta}{f_2} (uf_2 + xu + xf_2)$$

$$\simeq \dfrac{\beta}{f_2} (xu + f_2d). \qquad (3)$$

But $O_1b' = e \tan \dfrac{\beta}{2} = c \tan \dfrac{\alpha}{2}$

therefore $\beta \simeq \dfrac{c\alpha}{e}$.

Substitution for β in (3) gives $i \simeq \dfrac{c\alpha}{ef_2} (xu + f_2d). \qquad (4)$

If P is to be an anallactic point, α must be constant and P fixed as the position of L2 changes (i.e. as u and x change). Let the values of u and x when the staff is at infinity be u_∞ and x_∞ respectively.

Then $u = u_\infty + \Delta u$

and $x = x_\infty + \Delta x = x_\infty - \Delta u$,

because $(u + x) = d$ must be constant. It is necessary to see whether values for u_∞ and x_∞ can be found such that α is constant as x and u change, i being assumed constant, or, which amounts to the same thing, such that i is constant when x and u change, α being assumed constant.

Equation (4) becomes for S at infinity;

$$i_\infty \simeq \frac{c\alpha}{ef_2} (x_\infty u_\infty + f_2\, d). \tag{4a}$$

Thus $i - i_\infty \simeq \Delta i = \dfrac{c\alpha}{ef_2} (xu - x_\infty u_\infty).$

Therefore, $\Delta i = \dfrac{c\alpha}{ef_2} \left[(x_\infty - u_\infty)\Delta u - \Delta u^2 \right] = 0 \tag{5}$

Equation (5) expresses the condition for P to be an anallactic point. It is not possible to satisfy this equation given the physical nature of the telescope, so it is impossible for the internally-focusing telescope to be completely anallactic. However, suppose x_∞ and u_∞ are chosen so that $x_\infty = u_\infty$.

Then, $\Delta i = \dfrac{-c\alpha}{ef_2} (\Delta u)^2$ and $\dfrac{\Delta i}{i_\infty} = \dfrac{-(\Delta u)^2}{x_\infty u_\infty + f_2 d} \tag{6}$

and variation in i is of the second-order.

In practice of course, i is constant and α varies. Variation of i with α constant has been considered above solely to simplify the working. To relate Δi to $\Delta \alpha$, consider Figure 2.17.

$$\frac{\dfrac{\Delta i}{2}}{\dfrac{i_\infty}{2}} = \frac{b''b''_\infty}{b''_\infty 0_2} = \frac{b'b'_\infty}{b'_\infty 0_1} = \frac{\dfrac{\Delta \alpha}{2}}{\dfrac{a_\infty}{2}}.$$

Therefore, $\dfrac{\Delta i}{i_\infty} = \dfrac{\Delta \alpha}{\alpha_\infty}.$

Substitution for $\dfrac{\Delta i}{i_\infty}$ in (6) gives

$$\frac{\Delta \alpha}{\alpha_\infty} = \frac{-(\Delta u)^2}{x_\infty u_\infty + f_2 d}. \tag{7}$$

In Fig. 2.17, let $\dfrac{S_\infty}{2} - \dfrac{s}{2} = \dfrac{\Delta s}{2}$.

Then $\qquad \dfrac{\triangle D}{D_\infty} = \dfrac{\Delta s}{s_\infty} = \dfrac{\Delta \alpha}{\alpha_\infty} = \dfrac{-(\Delta u)^2}{x_\infty u_\infty + f_2 d}$ from (7).

For a telescope such as that on the Wild T2,

$f_1 = 118.6$ mm,
$-f_2 = -43.9$ mm,
$x_\infty = u_\infty = 37.7$ mm and
$F = 220.6$ mm (see section 2.1.7).

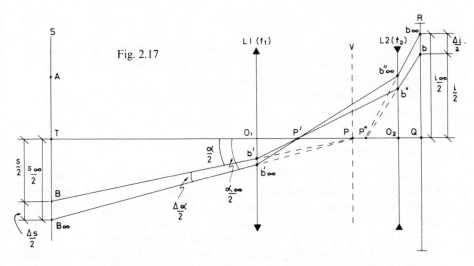

Fig. 2.17

The relation between D and $\triangle$D is shown in the following table.

D(m)	0.119	1.5	2	5	10	20	50	100	200	300	500	∞
$\triangle$D(mm)	∞	64.6	46.1	17.4	8.4	4.2	1.7	0.8	0.4	0.3	0.2	0

In practice, x_∞ is not made exactly equal to u_∞. This results in the error $\triangle$D being positive for long ranges and negative for short ranges. The error curve for a 210 mm focal length telescope is shown in Fig 2.18.

It can be seen that over the normal range for tacheometric work the error in the distance arising from the non-anallactic property of the telescope is less than 1 in 1000.

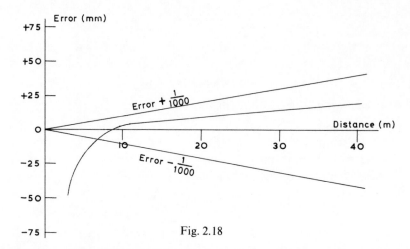

Fig. 2.18

Jackson, 1975 gives a résumé of the theory of anallactic imaging and Roy, 1979 gives the theory of anallactic telescopes which include a positive (converging) internal focusing lens.

An example of a modern internally focusing anallactic telescope is described in section 3.1.

2.1.10 Prisms used in surveying telescopes

For many years, surveyors were familiar with the use of telescopes which give inverted images. It was said that the inclusion of extra optical components solely to produce erect images was not necessary and was even undesirable because it was expensive and would result in both loss of brightness in the image and increased distortion caused by the extra refraction and reflection. In addition, the telescope would have to be longer to include the extra components which in turn would mean more weight, taller standards and less stability.

The development of compensators in levels led to the inclusion of additional components in the telescope. These often produced an upright image as well as compensation for residual tilts of the level. Improved design of the lenses, the use of light but strong alloys and more efficient coatings for reflecting surfaces were incorporated in the level telescope, with the result that there was no deterioration in the quality of the final image nor unacceptable increase in size or weight of the level. Users began to expect similar high-quality upright images in theodolite telescopes. The newer materials and techniques used in levels allowed the manufacturers to include them in redesigned theodolite telescopes with none of the former disadvantages.

Most recently, theodolite telescopes have been used not only to assist the observer's sighting of the target but also to collimate an amplitude-modulated infra-red signal for electromagnetic measurement of distance. Beam-splitter prisms are used to introduce the particular signal into the telescope for transmission through the objective to the reflector and at the same time to remove the reflected signal from the telescope for phase comparison.

It is possible in theory to use plane mirrors for producing an erect image and for beam-splitting but prisms are preferred in practice because of the relative difficulty in positioning a plane mirror with sufficient angular accuracy and rigidity.

2.1.10.1 Erecting prisms

Although the ray paths illustrated in Figs 2.19 and 2.20 are for prisms made of one type of glass only, it must be remembered that in a theodolite telescope a prism is usually inserted in a bundle of convergent rays. The angle of incidence at the refracting surfaces is zero for only that ray which passes along the optical axis of the telescope. All other rays will be deviated and the amount of the deviation will depend not only on the angle of incidence but also on the wavelength of the light. To keep chromatic aberration to a minimum, achromatic combinations of different glasses are often used.

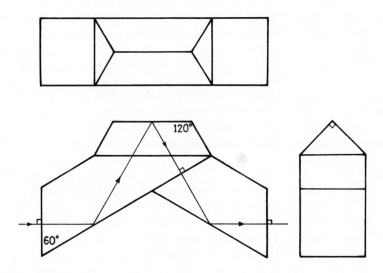

Fig. 2.19

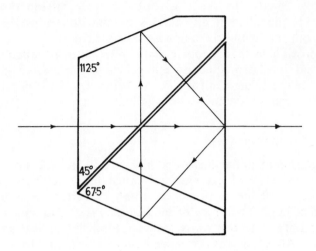

Fig. 2.20

At internal reflecting surfaces of the prisms, the convergent nature of the ray bundles may mean that for some rays the critical incident angle for total internal reflection will not be exceeded. For this reason and to reduce the amount of scattering, the reflecting surfaces of prisms are usually coated with reflecting material.

The simplest erecting prism is the Dove prism but it cannot be used in surveying telescopes because it will introduce astigmatism into a convergent bundle of rays. Porro prisms are also unsuitable as they introduce lateral displacement of the optical axis which would make a larger diameter of the telescope necessary. The *Abbé* (or *Brashear-Hastings*) prism is illustrated in Fig 2.19, where it is shown with a roof so that it reverses the image (left/right as seen in the telescope) as well as making it erect.

The *Pechan* (or *Schmidt*) prism is illustrated in Fig 2.20. The diagonal faces act as both transmitting and total internal reflecting surfaces, so reflective coatings cannot be used on them. The two parts of the prism must be separated so that total internal reflection and not transmission takes place at the centre. Thus the ray bundle must be sufficiently narrow so that all rays in it make angles of incidendence at the reflecting surfaces larger than the critical angle. Given these restrictions, the Pechan prism can be shorter than the Abbé prism and yet it has a useful optical path length.

2.1.10.2 The beam-splitter prism

This is illustrated in Fig 2.21. Two 90° prisms in conjunction form a cube. Before cementing, the hypotensal face of one of the prisms is given a coating which is highly reflective for infra-red radiation of about 0.9 μm wavelength but transmits radiation in the visible part of the spectrum. Any beam-splitter which discriminates between two spectral regions in this way is termed *dichroic*.

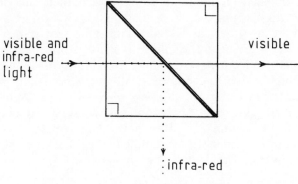

visible and
infra-red
light

visible

infra-red

Fig. 2.21

2.1.11 The Cassegrain telescope

In the classical form, this telescope consists of a primary parabolic mirror M_1 and a secondary hyperboloid mirror M_2 arranged as shown in Fig 2.22. Incident rays parallel to the common axis of symmetry of

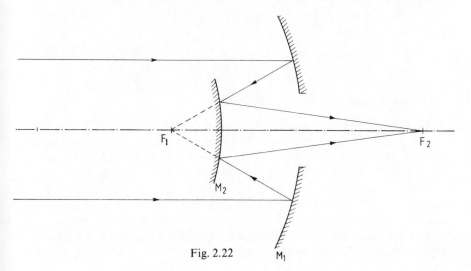

Fig. 2.22

the mirrors are reflected by the parabolic mirror to its Focus F_1. If M_2 also has a focus at F_1, it will reflect incident rays converging towards F_1 to its second focus F_2. The final image at F_2 will be free from spherical and chromatic aberration. There will however be coma and astigmatism for off-axis images. This can be corrected by suitable shaping of both M_1 and M_2.

The advantage of the Cassegrain system is that it provides a more compact telescope than the conventional theodolite telescope described in sections 2.1.7 and 2.1.9. The disadvantage is that the mirrors are expensive to manufacture because of their uniaxial nature. This latter disadvantage can be overcome to some extent by manufacturing a spherical mirror with an associated refracting surface, the latter giving to the spherical mirror some of the properties of a uniaxial reflector. When a combination of reflecting and refracting units is used, the system is called *catadioptric*. An example of a catadioptric system is the *Mangin mirror* where a doublet is used to correct for the chromatic aberration that would be introduced by a single refraction. Figure 2.23 illustrates a telescope with a Mangin mirror used as the primary. An example of the practical application of the Cassegrain telescope is described in section 3.1.2.

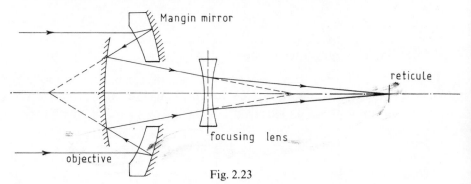

Fig. 2.23

2.2 Axis systems and kinematic design

A theodolite is an assembly of optical, mechanical and electronic components. These components must be located very accurately with respect to one another and should remain in those locations under normal field conditions. The mechanical design of the instrument should be carried out in accordance with *kinematic principles*. When one component is to be fixed in a specific spatial relationship to another, the location should be carried out only by those constraints which are necessary and sufficient to achieve the required spatial relationship. This *kinematic design* results in the smallest possible number of independent constraints, equal to the number of degrees of freedom between the components.

In general, to locate a particular component in space relative to a reference object, six degrees of freedom must be eliminated. These are three degrees of translational freedom and three of rotational freedom. The classic example (Fig 2.24) of kinematic design is in the location under gravity of a tripod having spherical feet, relative to a horizontal plane. One foot (A) is placed in a conical depression, the second foot (B) in a V-shaped groove with axis along AB and the third foot (C) on the plane. The conical depression eliminates two degrees of freedom (translations in the horizontal plane); the groove eliminates two more degrees of freedom (rotations about the vertical axis through A and about the horizontal axis AC); and the third foot C on the plane eliminates the third rotational degree of freedom (about the axis AB). Gravitational force constrains the tripod from moving in the third translational direction (vertically upwards). It can be seen that kinematic design results in constraints at points or along lines and not over surfaces.

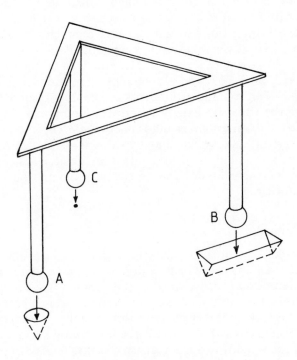

Fig. 2.24

If more than the minimum number of necessary constraints are used to locate a component, then either the redundant constraints are ineffective (and the design inefficient) or the component is under strain, possibly deformed and likely to introduce inaccuracies.

True kinematic design is often undesirable in practice. Theodolites must be portable and able to stand up to accidental knocks, vibrations and climatic changes. True kinematic design would be likely to result in an instrument too fragile to be used under those conditions. It is also undesirable to have kinematically designed components with a large number of moving parts. Contact at points and lines is often replaced in practice by contact over small surface areas. Pressures are thereby reduced and wear is likely to be less. When this concession to true kinematic design is made (and this is often the case in theodolite construction) the resultant design is said to be *semi-kinematic.*

Semi-kinematic design principles are applied not only to vertical axis systems, but to the location of prisms, lenses and circles in their mounts and to all other mechanical and optical components of theodolites. These mechanical and optical components are often located by means of pads so that the support is stress-free.

The trunnion axis of a theodolite in normal use is horizontal, or nearly so. A kinematically designed constraint of the axis under gravity with respect to the standards can therefore be achieved by two V-block bearings. For a cylindrical axis, contact in a V-block is along two lines. However, this true kinematic design is undesirable in practice. Wear will be unacceptable, and because the theodolite must be packed in a box and transported, an additional constraint is necessary to prevent the trunnion axis from falling out of the V-blocks. Usually, the trunnion axis is cylindrical, and located by three pads, spaced at 120° intervals around its circumference.

The vertical axis system has two functions; it has to support the weight of the alidade and maintain coincidence between the axis of rotation of the alidade (the vertical axis) and the centre of the horizontal circle.

2.2.1 Conical axis

In the older type of theodolite, the alidade axis (or inner centre) and the circle axis (or outer centre) are conical and adjacent, both lying inside a stationary sleeve fixed to the tribrach (see Fig 2.25). For the satisfactory operation of such a system it is necessary for the horizontal and conical surfaces to wear equally. If wear on the latter is greater, the axis wobbles and if the horizontal surfaces (which take most of the weight) wear more, then the axes tend to become wedged. With careful individual fitting, such a system can be made reliable for a

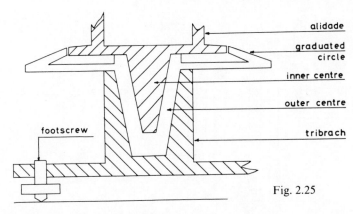

Fig. 2.25

short time. It can be seen that constraints are made over a large surface area and that the design is therefore non-kinematic.

2.2.2 Cylindrical axis

The conical axis system does not lend itself to mass-production methods and is inaccurate. Nearly all modern theodolites have hardened steel cylindrical axes, with a precision ball-race to carry the weight of the alidade. The centres are separated, the alidade axis is often inside the tribrach sleeve and the circle axis outside.

Figure 2.26 shows an early type of cylindrical axis system with separated centres. However, because of the large surface area used to constrain the alidade, this axis is not kinematically designed. A later, semi-kinematic design is illustrated in Fig 2.27. The ball-race not only bears the weight of the alidade, but also acts as a constraint of the upper end of the alidade axis. The constraint of the lower end is made over a very much smaller surface area than in the example illustrated in Fig 2.26. Manufacture of the later design does not produce the stresses

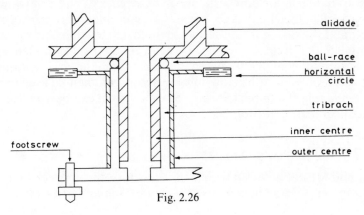

Fig. 2.26

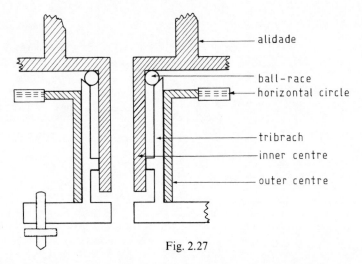

Fig. 2.27

set up by the extensive high-speed grinding necessary to produce the close fitting of the components in the earlier design, and wear is much less. The effects of the slow release of stresses introduced by high-speed grinding are discussed in section 5.5.

2.3 Controls for movement in azimuth

During the measurement of a horizontal angle, the circle generally stays fixed and the alidade is rotated. If the methods of repetition or reiteration are being used to measure the angle then it is also necessary to rotate the circle at certain times during the observing period. Three types of movement control are available.

2.3.1 Double-centre system

The alidade has a clamp (the upper plate clamp) and a slow-motion screw. When the clamp is on, the alidade is connected to the horizontal circle and the slow-motion screw can then be used to rotate the alidade relative to the circle. The circle also has a clamp (the lower plate clamp) and a slow-motion screw. When the clamp is on, the circle is connected to the tribrach, and the slow-motion screw can then be used to rotate the circle relative to the tribrach. Such a system can be used for either repetition or reiteration.

2.3.2 Repetition clamp system

The alidade has a clamp which connects it to the tribrach. The slow-motion screw can then be used to rotate the alidade relative to the tribrach. The horizontal circle is fitted with a 'repetition' clamp. In its usual position, this allows the circle to rest against the tribrach, but this clamp can be switched in order to clamp the circle to the alidade. This system can be used for either repetition or reiteration.

2.3.3 Circle-setting screw

In this system, the alidade controls are similar to those in the repetition clamp system. The horizontal circle normally rests against the tribrach, but it can be rotated by means of the circle-setting screw which engages a mesh connected to the circle. Thus there is no provision for carrying the circle round with the alidade so the repetition method cannot be used. Reiteration is possible.

2.4 Circle reading systems

Older theodolites with metal circles have graduations and angular values on the surface of the metal, either etched or engraved. Reading is by vernier or external micrometer, often with the aid of a simple magnifier or small microscope. The index mark is either the zero mark of the vernier scale or the wire in the micrometer and it should lie adjacent to the circle graduations so that parallax does not cause reading errors.

The introduction of a glass arc as the medium for carrying the angular graduations and values means that theodolites have become lighter and more stable. The glass is transparent, so the circles can be viewed indirectly by means of transmitted light reflected by mirrors and prisms either from the sky or from a small bulb powered by a D.C. electrical source. Optical micrometers, similarly illuminated, are often used to read the circles. The optical trains, optical micrometers and circles are completely enclosed within the theodolite casing which protects them from dust, rain and accidental damage. The circles and micrometers are often viewed through a small microscope with its eyepiece adjacent to the telescope eyepiece. Readings are made against an index mark carried on one of the plane surfaces in the optical train. The surface carrying the index mark and the plane of the circle must both lie in focal planes of the microscope so that the index mark and the circle graduations can be seen simultaneously and clearly by the observer and readings made without parallax errors.

Most recently, a few manufacturers have produced theodolites which have circles that are scanned by optoelectronic devices and not viewed by the observer. The glass arc is coded by a pattern of alternate opaque and transparent areas of glass. Incident light from an internal source falls upon the glass and photodetectors convert the light energy transmitted by or reflected from the pattern into electrical energy. The array of photodetectors acts as an index. As the alidade is rotated relative to the horizontal circle, the array of photodetectors rotates with it and scans a varying pattern of transmitted or reflected light. The corresponding change in the electrical output of the photodetectors is converted by a microprocessor into an equivalent angular value which is then displayed optoelectronically. A similar arrangement is used for the vertical circle.

There is often no zero mark or graduation on the circle. When this is the case, the circle reading system is said to be *incremental* to distinguish it from the *absolute* measurement relative to a zero graduation that is used in all theodolites read by eye. There is no commonly accepted method of coding circles. Digital (binary) codes, analogue (sinusoidal) codes, diffraction gratings and Moiré fringes derived from the circle code are some of the patterns used in recent optoelectronic circle scanning systems.

2.4.1 Optical reading systems

Suppose that the images in the reading microscope of the circle and the index mark (I) are as shown in Fig 2.28.

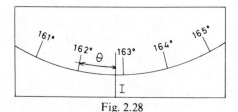

Fig. 2.28

The reading lies somewhere between 162° and 163°. To measure the angular value of θ, the amount by which the circle reading exceeds 162°, either an optical scale or an optical micrometer is used, although on some lower-order theodolites (e.g. Zeiss Oberkochen Th51) the reading is by estimation on the circle itself.

2.4.1.1 Optical scale

In this system, the face of the optical component which would normally carry the engraved index, carries instead an engraved scale whose length corresponds to the distance between two successive graduations of the circle. Figure 2.29 shows a scale running from 0′ to 60′ for a 1° graduation interval on the main scale. The index (the scale zero) is just past 22° and the full reading is 22° 02′.5, to the nearest 0′.5.

Fig. 2.29

2.4.1.2 Wedge micrometer

Referring to Fig 2.28, if the image of the circle can be displaced relative to the index so that the 162° graduation coincides with the index then a measure of the displacement introduced gives the value of θ. This displacement can be achieved by means of a travelling wedge prism.

In Fig.2.30, π_1 is the plane of the circle and G is a graduation on that circle. π_2 is the plane containing the index mark I. The image of G is at G'. The lenses are neglected for the sake of simplicity.

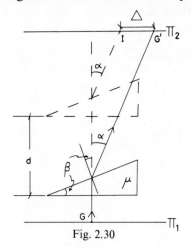

Fig. 2.30

If the wedge with refracting angle β and refractive index μ is moved through a distance d, parallel to the incident ray from G, the image G' will be displaced through a distance $\triangle$ onto the index I. Then, for the refraction at the prism face, Snell's law gives

$$\mu = \frac{\sin{(\alpha + \beta)}}{\sin{\beta}}$$

or, since $(\alpha + \beta)$ and β are small,

$$\mu \simeq \frac{\tan{(\alpha + \beta)}}{\tan{\beta}}$$

Expanding tan $(\alpha + \beta)$ and rearranging;

$$\mu \tan \beta \, (1 - \tan \alpha \tan \beta) \simeq \tan \alpha + \tan \beta$$

therefore, $(\mu - 1) \tan \beta \simeq \tan \alpha \, (1 + \mu \tan^2 \beta)$

but $\triangle = d \tan \alpha$,

therefore, $$\triangle \simeq \frac{d \, (\mu - 1) \tan \beta}{1 + \mu \tan^2 \beta}.$$

Therefore $\triangle$ is nearly proportional to d, and if d, β and μ are known, $\triangle$ can be calculated.

The displacement d is achieved by a mechanical linkage from a micrometer screw. This screw is also connected by a mechanical linkage to a micrometer scale. This scale is rotated against an index, the reading on the scale being directly proportional to $\triangle$, but in units of angular measurement. If the radius of the image of the circle is R, then the micrometer reading will be $\triangle / R = \theta$. It is necessary to see whether this approximation of $\triangle$ to the arc length s actually required is satisfactory.

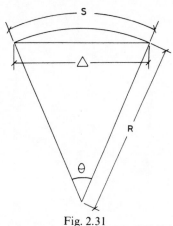

Fig. 2.31

Referring to Fig 2.31, $\theta = s / R$, but the value of θ taken is $\theta' = \triangle / R$. Therefore the error in θ is $\theta' - \theta = \delta\theta$, say where

$$\delta\theta = \frac{1}{R} \, (\triangle - s).$$

But $\triangle = 2R \sin (\theta/2)$ and $s = R\theta$.

Therefore, $$\triangle = 2R \cdot \left(\frac{\theta}{2} - \frac{1}{3!} \left(\frac{\theta}{2} \right)^3 + \ldots \right)$$

therefore,

$$(\triangle - s) \simeq - \frac{2R}{3!} \left(\frac{\theta}{2}\right)^3$$

therefore,

$$\delta\theta \simeq - \theta^3/24.$$

For a graduation interval of 20' (1200"):

$$\delta\theta'' \simeq - \frac{(1200)^3}{24 \, (\rho'')^2} \simeq 1.6'' \times 10^{-3}$$

Thus no appreciable error is introduced by the approximation.

2.4.1.3 Parallel-plate micrometer

In this case (Fig 2.32) the circle graduation G will have its image at G' when the parallel plate is normal to the incident ray from G. G' can be made to coincide with the index I by rotating the plate through an angle i, the resultant displacement being $\triangle$. If t is the thickness of the glass block which has a refractive index μ,

$$\sin i = \mu \sin r$$

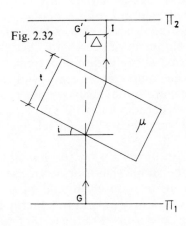

Fig. 2.32

Fig. 2.33

but, (from Fig. 2.33) $\triangle = BC \cos i$

But BC = CD − BD $= t \tan i - t \tan r.$
Therefore $\triangle = t (\tan i - \tan r) \cos i$

$$= t \left(\tan i - \frac{\sin i}{\mu \cos r}\right) \cos i$$

If i and r are sufficiently small, then the approximations i = tan i = sin i and cos r = cos i = 1 can be made so that

$$\Delta \simeq ti\,(1 - \frac{1}{\mu}).$$

However, in practice i is of the order of $\pm$ 10° and the approximations are not justified.

Thus
$$\Delta = t \left[\frac{\sin i}{\cos i} - \frac{\sin i}{\mu \left\{ 1 - \frac{\sin^2 i}{\mu^2} \right\}^{1/2}} \right] \cos i$$

which simplifies to give

$$\Delta = t \sin i\,[1 - (1 - \sin^2 i)^{1/2}\,(\mu^2 - \sin^2 i)^{-1/2}].$$

If the micrometer scale which gives Δ is to be linear, then the linkage between the micrometer knob and the parallel plate must be designed to take account of the variation of Δ with i.

2.4.2 Components of optoelectronic systems

The devices used for the illumination of the coded circles and for the detection of the transmitted or reflected light are *solid-state diodes*. The use of the description *solid-state* distinguishes the device from the vacuum diode. The invention and development of solid-state devices in the last few decades have led to the mass-production of extremely small, robust and inexpensive electronic components needing only a few volts for their operation. The incorporation of such components into theodolites has begun in the last two decades and will continue as the traditional optical and mechanical components become relatively more expensive to manufacture.

The solid-state diode depends for its operation on the *junction effect*. An atom of crystalline silicon, for example, has four electrons in the outer valency band. These electrons are loosely bound to the parent atom in the crystal lattice. At room temperature, thermal energy can be taken up by some of these electrons with the result that they may have sufficient energy to overcome the relatively weak forces that bind them to the parent atoms. When this happens, the electrons and the positively charged ions (the *holes*) are able to move at random through the lattice. If an external electric field is applied to the silicon crystal, the electrons will tend to move in one direction, the holes in the opposite direction and a small current will flow. Electrons removed from the valency band and able to flow through the lattice under the influence of a small, externally applied electric field are said to be in the *conduction band*. It is this ability of a crystal at room temperature to pass a small current that had led to the name *semiconductor*.

If very small amounts of certain impurities are added to the silicon, the numbers of electrons and holes can be greatly increased, giving a corresponding increase in conductivity. This deliberate addition of an impurity to an intrinsic crystalline element is known as *doping*. For example, trivalent indium added to tetravalent silicon acts as an acceptor of electrons from the conduction band and results in an excess of holes. The doped crystal is then called *p*-type material. On the other hand, pentavalent antimony is a donor of electrons. When a small amount of it is added to silicon, an excess of electrons in the conduction band is produced and the doped silicon is *n*-type material. When the intrinsic semiconductor element is doped in this way, the effects of the doping are very much greater than the corresponding properties of the intrinsic element itself.

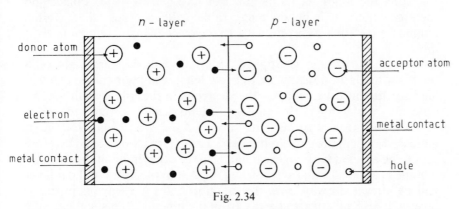

Fig. 2.34

Figure 2.34 shows a *p-n junction*. The donor and acceptor atoms are bound in the lattice, but the electrons and holes arising from the doping are free to move within the lattice if thermal or electrical energy is available. Some of the electrons in the conduction band of the *n*-layer will be attracted by the holes in the *p*-layer and will cross the junction, combining with the holes to form neutral atoms. Similarly, some of the holes in the *p*-layer will cross the junction and combine with electrons in the *n*-layer. The result of these movements will be the formation of a layer around the junction (Fig 2.35) within which the holes and conduction band electrons are depleted. This layer is the *depletion layer*. A *space charge* is created in this layer by the presence of oppositely charged donor and acceptor atoms uncompensated by equivalent numbers of electrons and holes. Ultimately, the *contact potential* arising from the space charge is sufficient to stop the drift of

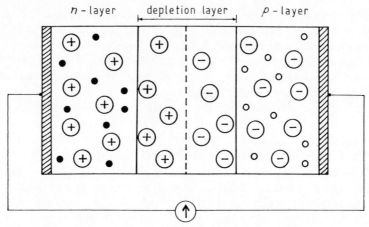

Fig. 2.35

electrons and holes across the junction. No nett flow takes place and therefore there will be no measurable voltage across the external connections. Such a device with two electrodes is a *semiconductor diode*.

If a small voltage is applied externally to the terminals of a semiconductor diode as shown in Fig 2.36a, the holes in the *p*-layer will tend to flow into the *n*-layer, the electrons in the *n*-layer will tend to flow into the *p*-layer, the contact potential will be reduced and a current will flow through the diode. As the external voltage is increased, so will the current increase. The external voltage is therefore said to be applied with *forward bias*. On the other hand, if the external voltage is applied with *reverse bias* (Fig 2.36b) the holes in the *p*-layer will be attracted away from the junction and the electrons in the *n*-layer will also be attracted away from the junction. This will have the effect of increasing the space charge in the depletion layer. The contact potential will be opposite to the applied voltage and only a few

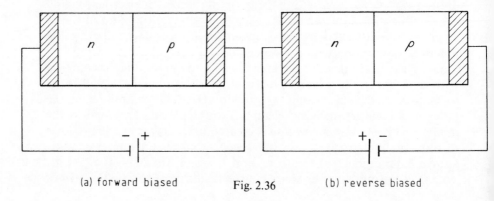

(a) forward biased   Fig. 2.36   (b) reverse biased transitions of holes and electrons under the influence of thermal energy will take place. Thus only a very small current will flow and this will not increase with an increase in the external voltage. The junction diode can therefore be used as a rectifier of alternating current.

2.4.2.1 The *p-i-n*, Schottky barrier and avalanche photodiodes

A *p-i-n* photodiode consists essentially of a layer of intrinsic (*i-*) material between *p-* and *n*-layers. If the depletion (*i-*) layer of a reverse-biased junction *p-i-n* diode is irradiated by light (Fig 2.37 after Ross, 1979) some absorption of the incident photons will take place and

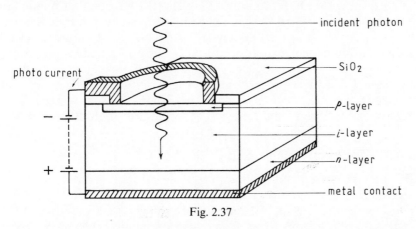

Fig. 2.37

some photons will pass through the *i*-layer. Some of the energy of the incident photons that are absorbed will be transformed to heat energy (causing vibration of the crystal lattice) and under certain conditions will also have the effect of raising electrons from the valency band of the intrinsic material to the conduction band. There will then be a reduction of the space charge and the small current that flows under reverse bias (sometimes called the *dark current*) will be increased. This increase in current is the *photocurrent* and it is proportional to the power of the incident radiation, assuming that absorption takes place only in the depletion layer.

For a silicon photodiode, the incident photons must have an energy equivalent to a radiation wavelength between about 400 nm (violet) and 1 μm (near infra-red) if their absorption is to result in increased numbers of electrons in the conduction band of the intrinsic semiconductor material. The electrons and holes freed by the radiation are captured immediately they leave the depletion layer, so the gain of the *p-i-n* photodiode is unity – it is non-amplifying. The photodiode acts as a detector of incident light within a certain band of wavelengths. A photocell is similar to a photodiode but it generates its own voltage, whereas the photodiode needs a reverse-biased voltage to function.

The *Schottky barrier* photodiode makes use of a metal semiconductor junction formed by a very thin film of gold evaporated onto *n*-doped silicon. The advantages of this combination over the silicon *p-i-n* photodiode are firstly, its response time is shorter and secondly, its sensitivity to the shorter wavelengths of the visible spectrum is higher. An anti-reflection coating is usually applied to the metallic layer to increase the amount of light which passes through it to the depletion layer.

Both the *p-i-n* photodiode and the Schottky barrier photodiode are non-amplifying. When the incident radiation to be detected is weak (as it is for a return EDM signal, for example) then an internal amplification of the current produced by the photodiode is necessary. An *avalanche* photodiode produces a gain in current as a result of electrons freed by the absorption of the radiation freeing additional electrons by impact ionisation. The reverse-biased voltage level determines the gain produced by the avalanche photodiode.

2.4.2.2 The electroluminescent diode (LED)

If a forward-biased voltage is applied to a *p-n* junction diode, electrons and holes will move into the depletion layer where they combine in pairs with the release of energy. Under certain conditions the transition of electrons from the conduction band to the valency band releases photons and the diode is then an electroluminescent diode (or LED – from light-emitting diode). This process of electroluminescence can be looked on as the opposite of the production of a photocurrent by irradiation of a photodiode but the processes are not reversible.

When gallium arsenide (GaAs) with silicon doping is used as the material of the semiconductor, the electroluminescent radiation is in the infra-red part of the spectrum (about 900 nm wavelength) and therefore invisible. It can however be detected by a silicon photodetector and the two solid-state devices are often used in short-range electromagnetic measurement of distance (EDM). The fact that the intensity of the electroluminescent radiation is directly proportional to the input current to the LED (up to about 40 mA) makes the amplitude modulation used for the distance measurement easy to produce.

If some of the arsenic in GaAs is replaced by phosphorus, the resultant mixed crystal of gallium arsenide phosphide (GaAsP) is also a semiconductor and emits red to yellow light, depending on the relative proportions of arsenic and phosphorus. The highest electroluminescent efficiency occurs at about 670 nm wavelength (red) and radiation bright enough to be seen in normal daylight can be obtained from only 10 mA of current. The radiation lies within a narrow cone of 22° angle.

In practice, to increase the efficiency of the electroluminescent output the GaAsP layer is based on a gallium phosphide (GaP) substratum (Fig 2.38 after Ross, 1979) so that the light emitted downwards is totally internally reflected at the face of the GaP crystal and ultimately emerges from the device.

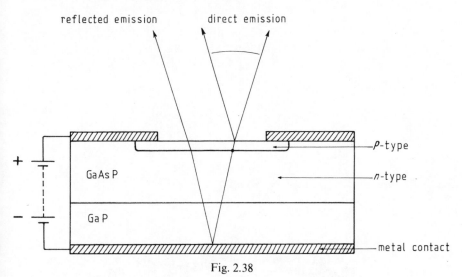

Fig. 2.38

2.4.2.3 The Gray code

The angular position of an index mark relative to a theodolite circle is a continuously varying quantity. When it is necessary to represent a continuous variable by coded binary digits, it is desirable that only one digit at a time changes as the variable changes by equal increments. Such a binary code is called a *progressive* code. The simplest progressive binary code is the Gray code. Figure 2.39 shows the Gray code and pure binary code representations of the decimal numbers 0–15 inclusive. Four digits are necessary to give the binary equivalents of these sixteen decimal numbers. The underlined Gray code or pure binary code digits are the digits that have changed in going from the preceding decimal number. It can be seen that after the first change, the Gray code digit 1 changes with every other increment of the decimal equivalent, the Gray code digit 2 changes with every fourth decimal increment, the digit 3 changes with every eighth, and so on, the nth Gray code digit changing once every 2^n increments of the decimal number. The Gray code is illustrated diagrammatically in Fig 2.40 which shows an arc of a circle divided sexagesimally and coded by a six-digit Gray code.

Decimal	Gray code	Pure binary
0	0 0 0 0	0 0 0 0
1	0 0 0 1	0 0 0 1
2	0 0 1 1	0 0 1 0
3	0 0 1 0	0 0 1 1
4	0 1 1 0	0 1 0 0
5	0 1 1 1	0 1 0 1
6	0 1 0 1	0 1 1 0
7	0 1 0 0	0 1 1 1
8	1 1 0 0	1 0 0 0
9	1 1 0 1	1 0 0 1
10	1 1 1 1	1 0 1 0
11	1 1 1 0	1 0 1 1
12	1 0 1 0	1 1 0 0
13	1 0 1 1	1 1 0 1
14	1 0 0 1	1 1 1 0
15	1 0 0 0	1 1 1 1
Digit no.	4 3 2 1	4 3 2 1

Fig. 2.39

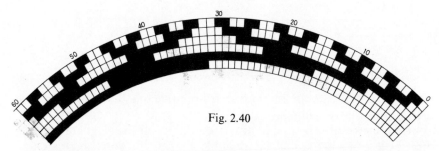

Fig. 2.40

To divide the circle into sectors, each of 10″ angular width, would necessitate 129 600 divisions around the circumference. For a circle of diameter 80 mm, each sector would have to be less than 2 μm wide at the circumference. Eighteen binary digits would be necessary to express the range 0°–360° in increments of 10″ uniquely. The construction and reliable optoelectronic scanning of such a circle with a linear array of 18 sensors, all within the dimensions of a theodolite, present practical difficulties which cannot yet be overcome economically. In section 3.4.4 some of the recent practical methods of coding and optoelectronic scanning of theodolite circles are described. None of these uses solely a simple unique code like the Gray code.

Gorham, 1976 describes binary coded circles and their application in the Zeiss (Oberkochen) Reg Elta 14.

2.4.2.4 Optoelectronic couplers

Figure 2.41 illustrates in schematic form a section of the circumference of a theodolite circle, coded with a three-digit binary code. The array of luminescent diodes, the coded circle and the array of photodetectors constitute an optoelectronic coupler.

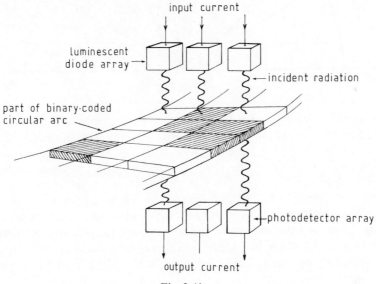

Fig. 2.41

To design an optoelectronic coupler suitable for detecting and measuring angular rotation in a theodolite, it is necessary to consider certain characteristics of the source of light, of the coded glass and of the photodetector and to ensure that the three components can be used together to produce a reliable and efficient measuring system.

The *radiation intensity* of the source and the *radiation sensitivity* of the detector depend on the angle of incidence. The intensity (and sensitivity) is generally defined in terms of polar co-ordinates, often in graphical form. Figure 2.42 illustrates an intensity distribution for a luminescent diode. The maximum intensity is at an angle of about 10° to the normal of the face of the diode. Factors which affect the distribution of the intensity are: centring of the wafer of semi-conductor material in its casing; the shape of the wafer; internal reflections within the crystal; the shape of external domes (of epoxy-resin for example), lenses and prisms and their positions in relation to

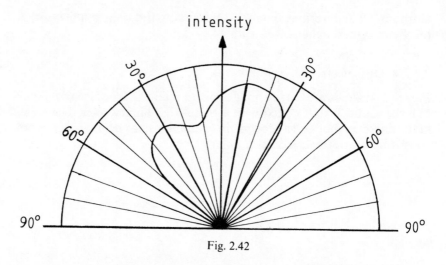

Fig. 2.42

the wafer, and certain electrical and thermal parameters. By a suitable arrangement of these factors, the radiation intensity of the source and the sensitivity of the detector can be matched, bearing in mind the loss of intensity caused by the light passing through the glass circle.

The *spectral sensitivity* of the photodetector must be compatible with the *spectral radiation* of the source. There will be an upper limit of wavelengths to which the detector responds. This upper limit corresponds to the minimum energy required to raise an electron from the valency band to the conduction band (section 2.4.2). The variation of spectral sensitivity, $S(\lambda)$, is illustrated in Fig 2.43 for a normal

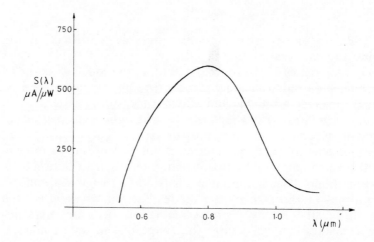

Fig. 2.43

silicon photodiode. Spectral sensitivity is expressed in units of μA output per 1 μW of incident radiation.

Dynamic characteristics of the photodetector are of great importance in an optoelectronic coupler designed to measure the rotation of a coded glass circle. The *switch-on time* is defined as the time taken for the output current to rise from zero to 90% of its full value and the *switch-off time* is the time taken for the output current to fall from 100% to 10% of its full value. If the input current to the luminescent diodes in Fig 2.41 is continuous and the diode and photodetector arrays are rotated relative to the circle, the photo-detectors will receive pulses of light which will have decreasing duration as the speed of rotation of the alidade is increased. Although the switch-on time might be as short as 10 ns for a silicon *p-i-n* photodiode, a total time of 10 μs is more typical of the minimum necessary for a photodetector and counter to register the transition of one coded section of the circle. If the circle is coded uniquely for each 10″ increment, then the coupler will just be able to maintain registration if the alidade is rotated once every $360 \times 60 \times 6 \times 10$ μs, i.e. once every 1.3 s. If the alidade is rotated more quickly, then registration of the rotation will be lost.

2.4.2.5 The controlling microprocessor

The electrical output from the optoelectonic devices is of course not immediately useful to the observer. The registration, transformation and display of data are carried out by a microprocessor, usually enclosed within the theodolite casing. Figure 2.44 illustrates the basic

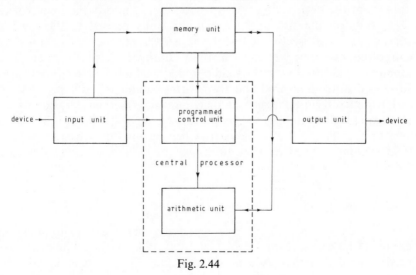

Fig. 2.44

structure of a microprocessor. Data enter the microprocessor through the input unit. These data might come from thumbswitches or a small keyboard, from an array of photodetectors or, in the case of a theodolite designed also for EDM, from a phase discrimination circuit. The programmed ROM (read-only memory) control unit accepts the incoming data and transformations take place in the arithmetic unit. Results are stored in the memory unit. The arithmetic unit and the memory unit are examples of RAMs (random access memories). Transformations that are commonly required by the operator are: conversion of binary coded circle readings to decimal values; referencing and indexing of vertical circle readings; and with EDM, conversion of measured phase differences to slope distance, correction for refractive index and reduction to horizontal and vertical components of the distance measured. After the transformations, certain data, often selected by the operator's manipulation of a keyboard, can be sent via the output unit to an external memory unit and to a luminescent diode or liquid crystal display. A detailed description of the microprocessor used in a specific instrument is given in section 3.4.4.6.

When the microprocessor controls the measurements and transforms the binary data to decimal digital displays, it is convenient to refer to it as the *controlling microprocessor*. It usually carries out simple computations (reduction of slope distances to the horizontal, for instance) but its primary function is to control the automatic measurement processes. The controlling microprocessor can then be distinguished from a *field computer* which is often an optional extra device supplied by some manufacturers. The field computer receives data from the controlling microprocessor and allows the operator to carry out, on the site, further computations related to the specific survey task. Intersections, resections, traverses and setting-out data can often be computed by the field computer. Usually, there is a memory associated with the field computer which allows the operator to record both measured and transformed data for later processing.

The controlling microprocessor is essential for the operation of the electronic theodolite, but the field computer is not, although it is often very convenient to use particularly when setting out. An example of a field computer used in a specific instrument is described in section 3.4.4.1.

2.4.2.6 The liquid crystal display (LCD)

The display of data to the operator is by either electroluminescent diodes (LEDs – section 2.4.2.2) or by liquid crystals. The latter are used generally when high levels of ambient light are expected and most electronic survey instruments use this form of display. When used at night, however, the LCD must be independently illuminated whereas the LED is luminescent.

Certain organic materials take up a physical form midway between that of a liquid and a crystal and remain in this state over a wide temperature range. The long molecules of an organic liquid crystal tend to line-up with their longitudinal axes parallel and the crystal is transparent. If a thin layer of the liquid crystal is contained between parallel electrodes and if a small voltage is applied to the electrodes, the molecular regularity will be disturbed. If the voltage level is sufficiently high (about 1 V per 1 μm thickness of crystal) turbulence will be induced by the current within the liquid crystal which will become opaque. Removal of the applied voltage results in the return of the liquid crystal to its ordered state within about 0.1 s. Levi, 1980 deals with the properties and characteristics of different organic liquid crystal compounds.

An array of seven units in the form of a figure 8 (⊟) allows for the display of any one of the decimal digits 0–9 inclusive. For alphanumerical characters, a 7 × 5 matrix of elements allows for the display of any one of the letters A–Z inclusive or any one of the ten decimal digits. It is desirable to keep the amount of data displayed at any one time to a minimum. The numerical display of a circle reading to one centesimal second requires at least seven digits and a decimal point, corresponding to 50 separate units. If a ten-character alphanumeric display is required, 350 separate liquid crystals or electroluminescent diodes will be needed. Each element of the array must be defined uniquely by the circuitry and within the memory of the controlling microprocessor. Large displays take up space in the theodolite and are undesirable and unnecessary.

2.5 The spirit level

This is essentially a glass vial with its interior surface ground to the required radius and almost filled with a liquid which has a low freezing point and a low coefficient of viscosity. Ether is suitable, with a freezing point of $-114°$ C and a coefficient of viscosity of 0.267×10^{-3} Ns/m^2 at 10° C. (Water has a coefficient of viscosity of 1.304×10^{-3} Ns/m^2 at 10° C.)

Figure 2.45 shows a longitudinal section containing C, the highest point of the vial.

The tangent at C in the longitudinal section is the principal tangent of the level. Sections through the vial perpendicular to the principal tangent are circular, varying from a minimum radius at the ends A and B to a maximum at C.

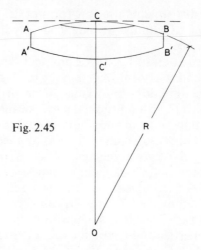

Fig. 2.45

The air-bubble will always come to rest with its centre at C, the highest point on the inside surface of the vial. The tangent plane at this highest point will be horizontal and a line through the centre of curvature 0 and the point of tangency will be vertical.

2.5.1 Displacement of the bubble

Suppose the principal tangent is displaced through a given angle α. Then the apparent displacement CC′ (Fig 2.46) of the centre of the bubble measured against the vial is proportional to the radius of curvature of the vial, and to α.

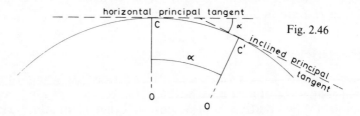

Fig. 2.46

In order to measure the apparent displacement of the bubble, reference marks, generally at 2 mm intervals are engraved on the outer surface of the vial, symmetrical with respect to C, but starting a few millimetres from C (see Fig 2.47).

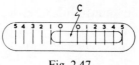

Fig. 2.47

These provide a reference for recording the position of the ends of the bubble; in Fig 2.47, the left-hand end is at graduation no. 1 and the right-hand end is at graduation 5. Hence, the highest point of the vial is $(5-1)/2$ i.e. 2 divisions to the right of the centre C. Therefore (for a 2 mm graduation) the centre of the bubble has been apparently displaced 4 mm towards the right. The value of α required to displace the centre of the bubble through 1 division is the 'bubble value', v. For a 2 mm division, $v = [2/R(mm)]$ radians or $v'' = [2/R(mm)]\rho''$. For a 25 m radius vial, v is about 20″. Generally the bubble value is engraved on the vial. Methods of determining bubble values are described in sections 4.7.1.1.

2.5.2 Motion of the bubble

If a vial with a stationary bubble is suddenly tilted through a small angle α, the bubble moves initially quite quickly, overshoots the final rest position, oscillates with decreasing amplitude and finally comes to rest in a displaced position relative to the vial. The time taken (T) for the bubble (in a given liquid) to come to rest can be expressed by the empirical equation $T = KR\,\alpha^{1/3}d^{-1/2}$ (Goulier) where R is the radius of curvature, d is the length of the bubble and K is constant for a given liquid at a given temperature. The greater the viscosity of the liquid, the longer it takes to come to rest, and the less obvious are any displacements.

Modern theodolites and levels are more compact than the older models and the bubbles therefore are shorter, leading to a longer settling-down time.

2.5.3 Bubble reading systems

The only bubble to be read by direct reading is the plate level on theodolites. This is generally at a convenient height and directly in front of the observer and can be viewed comfortably without parallax between the bubble and the graduations.

2.5.3.1 Inclined mirror

The altitude level of a theodolite is normally too high to be read directly in comfort and is at one side of the theodolite. One method of viewing the bubble is by an inclined plane mirror (Fig 2.48).

Suppose the bubble ACB is centred between the two graduations a and b. The image of the bubble A′C′B′ must appear symmetrically between the images of a′ and b′ of graduations a and b respectively. This relation will hold only if, for a given inclination α of the mirror, the eye

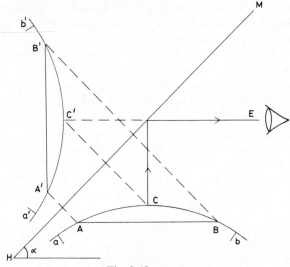

Fig. 2.48

is placed on the line EC′ normal to the chord A′B′. In such a case, the angles subtended at E by A′a′ and B′b′ are equal. If the view-point is away from the line EC′, then for the given value of α the bubble will appear off-centre. Generally the inclination α is maintained constant by a mechanical stop and the bubble is viewed from a point near the telescope eyepiece. It is not certain that the correct viewpoint will be used by all observers so this method is unsatisfactory except for lower-order theodolites.

2.5.3.2 Prism reader

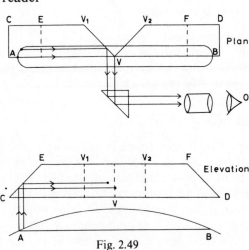

Fig. 2.49

Figure 2.49 shows a prism block CEFD situated above (or below) the vial with its plane face CD parallel to the principal tangent of the vial. The block covers only one half of the width of the vial. A vertical ray from the end A of the bubble on the longitudinal axis enters the prism block normally, meets the face CE inclined at 45°, is reflected horizontally onto the 45° face $V_1 V$, leaves the block normal to the exit face and passes via a 45° prism and a magnifier to the observer's eye at 0. A symmetrical path is followed by a vertical ray from B so images of A and B are superimposed. Rays adjacent to those from A and B follow similar paths so that the magnified images of the two half-ends of the bubble are adjacent. When the bubble is central, the image is as shown in fig 2.50a with the images of A and B at A′ and B′ respectively.

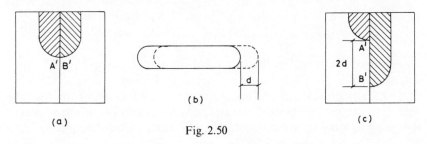

(a)

(b)

(c)

Fig. 2.50

Displacement of the centre of the bubble through a distance d (Fig 2.50b) results in each image being displaced by d, giving a relative displacement of 2d because movements of the two images are in opposite directions (Fig 2.50c). This then gives a 2 × increase in sensitivity (equivalent to a bubble of twice the radius but without the increase in time taken for the bubble to come to rest).

A further increase in accuracy over the direct – viewing and mirror – viewing systems is achieved by the coincidence setting; it is more accurate than placing two ends of a bubble symmetrically between two graduations.

Finally, the accuracy is increased still further by viewing the split-image through a microscope giving a magnification of about × 2.

A bubble viewed directly can be centred to an accuracy of about 0.4 mm with 2 mm graduations. The same bubble viewed by a coincidence reading system can be centred to about ten times this accuracy.

2.5.4 The spherical level

This level is used to set up a theodolite or target in approximately the correct attitude. The upper face is spherical with a radius of curvature of about 0.5 m. It is therefore not very sensitive but the time taken for the bubble to come to rest is very small and generally inappreciable.

The bubble casing usually has three adjusting screws in its base. The adjustment is described in section 9.1.1.

2.5.5 Level vial mountings

The glass vial is set in plaster-of-Paris inside a metal shield (Fig 2.51).

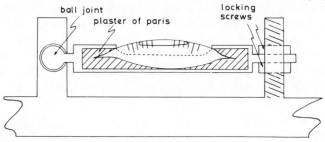

Fig. 2.51

One end of the shield is held in a ball joint and the other end has a threaded pillar through it, on which it is held fixed by a pair of locking screws (capstan screws). Adjustment of the level in order to place the principal tangent perpendicular to the vertical axis of the theodolite is made by raising or lowering this end of the case on the pillar.

2.6 Centring systems

The vertical axis, when vertical, must pass through the ground mark. There are three different methods of achieving the centring; by plumb-bob, optical plummet or centring rod. Levelling and centring are two operations carried out each time the theodolite is set up and they are generally carried out together since one often affects the other. Lateral adjustment necessary for centring can take place either above or below the footscrews. If it is below the footscrews (by moving the base-plate or trivet-stage on the tripod head, e.g. Wild T2) then movement will affect the levelling appreciably unless the tripod head is horizontal to within the accuracy required of the levelling or unless the movement is non-rotational. (Fig 2.52a) If centring takes place above the footscrews (e.g. Vickers Tavistock II) it will affect the levelling only very little. (Fig 2.52b)

Instruments making use of centring below the footscrews take longer to set up when using the optical plummet but are more compact than instruments using centring above the footscrews.

2.6.1 Centring by plumb-bob

If the plumb-bob is suspended from a point on the vertical axis and the

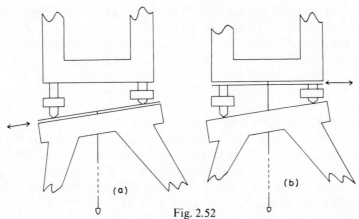

Fig. 2.52

theodolite is centred first, then levelled, the point of suspension will not move vertically but along an arc, thereby causing a centring error. Suppose (Fig 2.53a) footscrew C is turned.

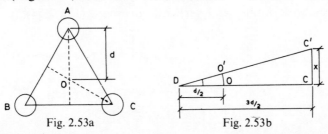

Fig. 2.53a   Fig. 2.53b

Rotation will take place about the line joining the centres of the other two footscrews A and B. Taking a vertical section through C and O (Fig 2.53b) suppose footscrew C is adjusted through a distance x. The point of suspension will move from O to O'.

The horizontal component of the arc OO' is $e = \dfrac{d}{2}(1 - \cos \alpha)$. Therefore, the miscentring e is given by:

$$e = \frac{d}{2}\left(1 - 1 + \frac{\alpha^2}{2!} - \frac{\alpha^4}{4!} + \ldots\right)$$

$$\simeq \frac{d}{2} \cdot \frac{\alpha^2}{2!}$$

But $\alpha \simeq \dfrac{2x}{3d}$

therefore, $e \simeq \dfrac{d}{2}\left(\dfrac{2x}{3d}\right)^2 \cdot \dfrac{1}{2!} = \dfrac{x^2}{9d}$

Suppose x = 10 mm and d = 50 mm

Then error $e \simeq \dfrac{10^2}{450} \simeq 0.2$ mm

Thus the error introduced is within the accuracy obtainable with a plumb-bob (normally about 1–2 mm).

On some theodolites (e.g. Wild) the plumb-bob is suspended from the lower end of the bolt which is used to clamp the trivet stage to the top of the tripod. The bolt is a few centimetres in length, so if the tripod head is off-level, the bolt will not be vertical and the point of suspension of the plum-bob will be displaced from the vertical axis of the theodolite, thereby causing a centring error. If the dislevelment of the tripod head is 3° and the point of suspension of the plumb-bob is 40 mm below the trivet, the centring error will be about 2 mm. Therefore it is necessary to ensure that the tripod for such theodolites is set up so that its head is approximately horizontal. The spherical level set in the top surface of the tripod head should be used for this setting-up.

2.6.2 Centring by optical plummet

The optical plummet consists essentially of a 45° prism with the two 90° faces horizontally and vertically disposed (Fig 2.54).

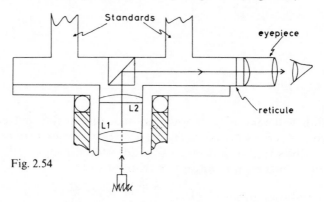

Fig. 2.54

A slight magnification of the station mark is usually present in the eyepiece. The intermediate image is formed at the reticule by the fixed lenses L1 and L2. The optical plummet is usually in the alidade of the theodolite. If centring is above the footscrews, generally no re-levelling is necessary after centring. If however, centring is below the footscrews, re-levelling will in general be necessary after centring. Then slight re-centring will probably be found necessary.

It is advisable to use the plumb-bob to obtain approximate centring before using the optical plummet.

2.6.3 Centring by centring rod

In this system, (Fig 2.55) a telescopic rod is placed with its top linked to

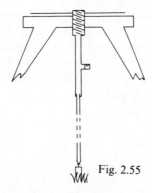

Fig. 2.55

the tripod and its base point placed on the station. The tripod legs are telescopic to facilitate centring and levelling.

When the rod is vertical (indicated by a spherical bubble attached to the rod) the screw thread attached to its head can accept the theodolite which will then be centred. The tripod head must be approximately horizontal so that levelling can be carried out using the footscrews. The rod usually is graduated so that the 'height of instrument' is obtained directly.

2.6.4 Comparison of centring systems

The plumb-bob is simple, cheap and accurate to about 1–2 mm. It is difficult to use in windy conditions, but can give an approximate setting before using the optical plummet.

The optical plummet can go out of adjustment (see section 4.6) and when used with centring below the footscrews, setting-up takes a little longer. On the other hand, theodolites with above the footscrews centring are taller, heavier and slightly more complicated. The accuracy is about 1 mm.

The centring rod is an additional piece of equipment for the surveyor to carry, but is very light. It is particularly useful when setting-up on steep slopes, and is generally quicker. Again the accuracy is about 1 mm.

2.6.5 Constrained centring

In many survey operations it is often necessary to replace one piece of equipment centred above a ground mark by another. For example in carrying out a traverse, it is necessary to interchange the theodolite and targets. Obviously, work is going to be speeded up and made more accurate if the theodolite for example can be removed from the tripod and the target put in its place and constrained so that its centre lies on the line previously occupied by the vertical axis of the theodolite. If the original centring of the theodolite over the ground mark was inaccurate, then the target will have a similar eccentricity and an angular error will not be propagated through the traverse.

Constrained centring systems fall into two main groups which are illustrated in Figs 2.56 a and b.

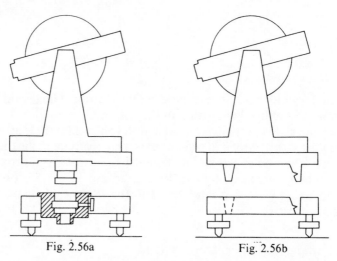

Fig. 2.56a   Fig. 2.56b

In the first group (systems with a stub axis and lateral clamp) the constraint is imposed by either a cylindrical stub axis or a conical stub axis fitting into a similar sleeve and held against the sleeve by a lateral clamp. The German 34 mm standard stub axis systems are present on Zeiss equipment, and some Vickers equipment make use of a short conical stub. Figure 2.56a illustrates constraint by cylindrical stub.

The second group are self-locating in their constraints and an example is shown in Fig 2.56b. This illustrates a system whereby three studs at 120° spacing are located in similar slots on the tribrach, clamping generally being effected by means of a sprung plate locating the notches in the studs.

The accuracies of different systems are described in section 5.6.

2.7 Vertical circle indexing and tilt sensor

The reading of the vertical circle can be made in the same way as that of the horizontal circle: by direct reading; by optical scale or micrometer; or by electronic methods. In each case, it is necessary to *index* the circle relative to the direction of gravitational force so that the reading is zero (or 90° exactly) when the line of collimation is horizontal. This indexing is conventionally carried out by using an altitude level and setting-screw as described in section 1.3 and Fig 1.4. The adjustment of the altitude level index is described in section 4.4.

Automatic indexing is often used in modern theodolites. For those theodolites which have optical methods for reading the circles, automatic indexing leads to quicker observations. Inexperienced observers often forget to set the altitude level index before reading the vertical circle, so automatic indexing sometimes leads to improved accuracy and fewer blunders. The automatic index is often referred to as a compensator because it is looked on as a device for compensating for the residual tilts of the vertical axis, but this nomenclature is incorrect. It has probably arisen from the use of the word 'compensator' to describe the device in a level where it is used to compensate for residual tilts of the standing axis of the level. Nevertheless, the use of the word 'compensator' is widespread and generally accepted as an alternative name for the automatic index of the vertical circle in a theodolite.

Two disadvantages of automatic indices are often quoted. The first is that they are more sensitive to the effects of vibration arising from wind, heavy machinery or traffic than the altitude level index. The second is that they tend to 'stick'. These criticisms are not always justified. In the first place, it is often quicker to take a few readings to obtain a mean value with a vibrating automatic index than it is to set an altitude level under similar conditions. In the second place, automatic indices are remarkably robust and stable, and the criticism that they tend to 'stick' is often made as the result of trying to take a reading outside the working range of the automatic index. This working range is normally a few minutes of arc. If the vertical axis is inclined to the vertical by more than this amount, most theodolites exhibit a visible warning in the field of view of the circle-reading microscope.

The automatic index is entirely enclosed within the theodolite casing, so it is less likely to go out of adjustment as a result of accidental damage than is an altitude level index which is often external to the theodolite casing. Adjustment of the automatic index is described in section 4.4.

There are two main methods of providing the automatic index for theodolites which use optical circle reading. Each makes use of the gravitational force. In the first method, a pendulum is used to define the vertical direction and in the second method, the surface of a liquid is used to define a horizontal plane. These two methods can also be used in theodolites which make use of optoelectronic scanning of the circles when the scanning is made through the index. The two methods are described in sections 2.7.1 and 2.7.2.

Another method of providing an index for the vertical circle of a theodolite which makes use of optoelectronic scanning is one where the surface of a liquid is used to define a horizontal plane and the tilt of the vertical axis relative to the normal to the surface of the liquid is measured optoelectronically. The controlling microprocessor is used to compute the corresponding corrections to the circle readings. Such a device is called a *tilt sensor*. It can be used not only to detect and measure the tilt of the vertical axis in the direction of the line of sight (for automatic numerical indexing of the vertical circle) but also to detect and measure the tilt of the vertical axis in a direction normal to the line of sight and parallel to the trunnion axis. This tilt measurement can be used by the controlling microprocessor to correct horizontal circle readings according to the questions described in section 4.7.1. Principles of a tilt sensor are described in section 2.7.3 and an application is described in section 3.7.2.

2.7.1 Liquid index

Use can be made of the fact that a small area of the free surface of a liquid assumes a horizontal plane under the influence of gravity alone.

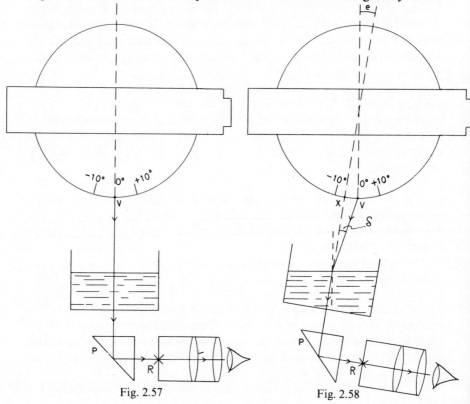

Fig. 2.57   Fig. 2.58

Suppose (Fig 2.57) that the telescope is horizontal and that the vertical axis is vertical. A ray from V on the circle will, after passing through the liquid, continue undeviated to the prism P and will fall on the optical index R of the circle reading microscope. Thus the reading will be correct at 0°. Now suppose (Fig 2.58) that the vertical axis is inclined at an angle e to the vertical in the plane of the circle, but the telescope remains horizontal. If the liquid were not present, the image formed at R (the index of the reading microscope) would be that of X and the reading would be incorrect. However, it is possible with a suitable liquid to refract the light rays from the circle so that the image of V is formed at R as if no tilt e were present.

The requirement for this to happen is that the deviation δ produced by the liquid is given by $\delta \simeq (\mu - 1)\,e$ where μ is the refractive index of the liquid. This can be seen by applying Snell's law to the refraction at the upper face of the liquid and taking $\sin \delta \simeq \delta$ etc.

If the radius of the vertical circle is r and the distance between the circle and the upper surface of the liquid is D, then the refractive index μ required to allow for the tilt can be seen to be $\mu \simeq 1 + r/D$ since $r e = D\,\delta = \mathrm{arc}\ XV$.

Other practical applications and developments of this principle are given by Clark, 1967.

2.7.2 Pendulum index

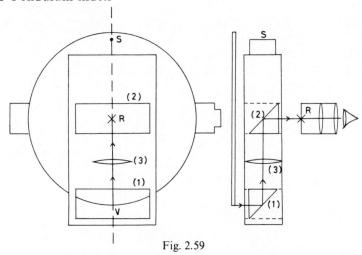

Fig. 2.59

In Fig 2.59, suppose that the telescope is horizontal and the vertical axis vertical. Then, with a reading system consisting of prisms (1) and (2) and lens (3) attached to a suspended plate, an image of the zero of the circle (V) will be formed at the reticule R on the transit axis of the theodolite. S is the point of suspension of the index device from one of the standards.

Suppose now the vertical axis is displaced through an angle e in the plane of the circle but that the telescope remains horizontal. Under the influence of gravity alone the suspended plate will hang vertically from P, the centre of instantaneous rotation (Fig 2.60).

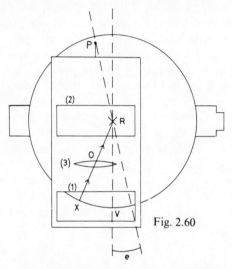

Fig. 2.60

Thus the circle reading at the reticule R will be at X, along the line joining R to the optical centre, O, of the lens (3). This reading is not the correct reading which should be at V. In order to obtain this reading, suppose the suspended plate is constrained to swing only through an angle α (Fig 2.61) such that O is on the line RV.

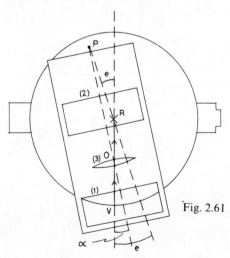

Fig. 2.61

The value of angle α required is given by $\dfrac{\sin \alpha}{\sin e} \simeq \dfrac{\alpha}{e} \simeq \dfrac{PR}{OP}$ (from triangle PRO).

The constraint on the suspended plate is imposed by the flexural rigidity of the suspension. By a suitable selection of values for the modulus of elasticity and moment of inertia of the X-section of the suspension, and by a design of distances PR, OP etc., and the weights of the components, the above relation $\alpha \simeq PR.e/OP$ can hold for values of e up to a few minutes. This type of index is considered in more detail in section 7.10.3, but in relation to compensators in automatic levels.

2.7.3 Tilt sensor

In electronic theodolites and tacheometers, the residual inclination of the vertical axis to the vertical can be measured by optoelectronic methods and the values then used by the controlling microprocessor to correct automatically the circle registrations.

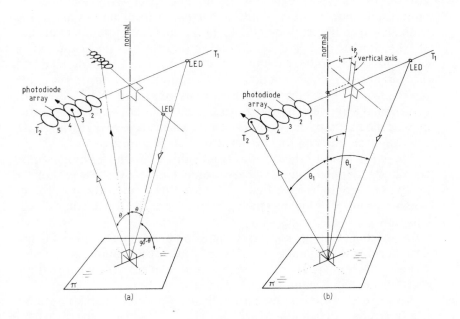

Fig. 2.62

The principle of the tilt sensor is illustrated in Fig 2.62a and b. The reference vertical direction is the normal to the plane surface π of a liquid at rest. In Fig 2.62a, T_1T_2 represents a line parallel to a horizontal trunnion axis which is assumed to be perpendicular to the vertical axis, which in turn is therefore vertical and normal to the surface of the liquid. Radiation from an electroluminescent diode (LED) on T_1T_2 is focused to fall on an array of photodiodes after reflection from the surface of the liquid. The focusing lenses are not shown and only the central ray of the bundle is illustrated. At the reflecting surface, the angle of incidence θ causes the reflected ray to fall on the central (no. 3) photodiode which emits a current. This signal is taken by the controlling microprocessor to indicate that the trunnion axis parallel to T_1T_2 is horizontal. A similar configuration of LED and photodiodes perpendicular to T_1T_2 results in the generation of a current from another central photodiode. These two outputs indicate to the controlling microprocessor that the vertical axis is vertical and the circle registration values are correct.

Now suppose that there is an inclination i of the vertical axis to the normal to the surface of the liquid (i.e. the vertical). Suppose that this inclination can be resolved into a component i_t in the direction of the trunnion axis and a component i_p in the direction of the telescope. The ray from the LED on T_1T_2 makes an angle of incidence θ_1 with the normal to the surface of the liquid and after reflection falls on photodiode no. 5 which emits a current. From the design of the geometrical configuration of the LED and focusing system, the surface of the liquid and the individual photodiodes, the value of i_t corresponding to output from photodiode no. 5 on T_1T_2 is defined and the microprocessor programmed to compute corrections to the circle registrations. These corrections are discussed in section 4.7. A similar, orthogonal arrangement allows for the determination of i_p and for the appropriate automatic correction to the vertical circle registration. In this way, the indexing of the vertical circle is automatic.

There is a limit to the tilt that can be detected which is imposed by the physical dimensions of the theodolite. Generally, the extent of the array of photodiodes is sufficient to detect tilts of up to about 2'. Either audible or visible warnings to the operator are given if the tilt exceeds this limit.

In practice, a single source of radiation and discrete sensors is not used because it does not give sufficiently precise discrimination of tilts. Instead, use is made of optoelectronic sensing methods similar to those used for the circles and described in section 3.4.4.

An example of the practical application of the principles outlined above is given in section 3.7.2.

2.8 Circle referencing

Vertical circles with incremental coding (as opposed to those with absolute coding) must be *referenced* to the vertical axis of the theodolite so that when the line of sight is at right angles (or parallel) to the vertical axis, the datum for vertical angle measurement is defined. For all vertical circles with incremental coding, this reference is lost when the power supply to the theodolite is switched off. It is therefore necessary to reference the circle each time the theodolite is set up and switched on.

In the case of a horizontal circle with incremental coding, it is not necessary to make the reference to a fixed direction, although it is often desirable for practical reasons to do so. Some electronic theodolites define the datum of the horizontal circle as being coincident with the direction in which the telescope happens to be pointing when the instrument is switched on. The datum can then be redefined at any time by an appropriate instruction to the controlling microprocessor via the keyboard or switches. The manufacturer's handbook gives the instructions for referencing the vertical circle of a particular electronic theodolite. The principle of the method is that when a certain coded mark on the vertical circle passes a stationary referencing sensor on a standard, a pulse is triggered which starts the counting of the coded marks on the circle. There is usually some visible or audible warning if the referencing has not been carried out.

It should be noted that referencing is made within the theodolite and it takes no account of external factors such as the direction of gravity or any other particular direction. It refers the zero of the vertical circle to the vertical axis and in theory can be made with the theodolite in any attitude. An automatic index or tilt sensor will be necessary to relate the reference to the vertical as described in sections 2.7.1, 2.7.2 and 3.7.2.

Circles which have absolute coding (including those read optically) must also be referenced, but this is part of the constructional process. It is carried out in the factory during assembly and is not part of the surveyor's normal field procedure. This type of referencing is not lost when the power supply is switched off.

3 Features of modern theodolites

In this chapter, certain features of modern instruments are described in order to illustrate how the principles described in the preceding chapter are applied.

3.1 Telescopes

The traditional internally focusing theodolite telescope has recently been redesigned to serve both as a sighting device for angular measurements and as a collimator for infra-red or laser EDM. This redesign has not resulted in a uniform approach by the major manufacturers. Catadioptric systems and lens/plane mirror combinations have both been used successfully in the design of telescopes that are short enough to be transited, long enough to contain the extra components and give the high magnification, and yet retain the balance and stability necessary for accurate angular measurement.

3.1.1 The telescope of the Wild T1 Micrometer Theodolite

This is a modern example of the conventional internally-focusing theodolite telescope and is illustrated in Fig 3.1. It is anallactic, with a multiplying constant of 100. The objective is an anastigmatic triplet

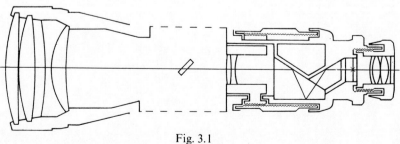

Fig. 3.1

with an aperture of diameter 42 mm. The focusing lens is translated by rotating a sleeve around the telescope barrel situated between the eyepiece and trunnion axis. This focusing sleeve has both coarse and fine motion. In Fig 3.1 the focusing lens is shown in the infinity focus position.

The shortest focusing distance is 1.7 m, but three auxiliary lenses are available which can be attached to the objective to reduce this shortest focusing distance. The focusing distances covered by the three auxiliary lenses are: 0.88–1.72 m; 0.63–0.92 m; and 0.50–0.65 m. Repeated accurate location of an external lens is difficult, so it should never be moved during observations. When directions to a near and a far target have to be observed in combination, it is better to take face-left and face-right readings to the near target before removing the auxiliary lens to sight the far target. Collimation errors are thereby reduced and not increased as they would be if the lens were repeatedly removed and replaced during rounds of observations.

The Abbé prism (section 2.1.10.1) gives an upright image, and with the standard eyepiece the magnification is 30×. Lower magnifications of 26× and 19× which can give higher observing accuracies when haze and shimmer are present can be obtained by using other optional eyepieces. For laboratory work, an eyepiece giving a 42× magnification is available. At this magnification, the deterioration of image quality at the edge of the field is very slight and does not adversely affect measurements. The field of view is just over 1.5°.

3.1.2 The telescope of the Hewlett-Packard 3820A Electronic Total Station

The HP 3820A is an electronic tacheometer. The telescope (illustrated in Fig 3.2) is used both as the sighting device for measurement of angles

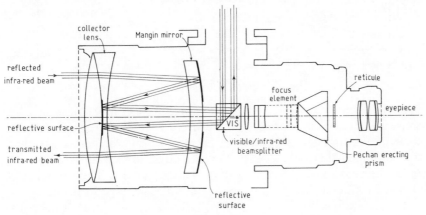

Fig. 3.2

and as the transmitter and receiver of the infra-red signal for EDM. The conventional anastigmatic triplet for the objective which is used in almost all theodolite telescopes has disadvantages when infra-red as well as visible light must be refracted. If the range of the EDM is to be useful (say up to a few kilometres) then the objective aperture must be larger than is normally the case so that more of the weak return signal is collected for measurement. This means that the aberrations of the objective are more difficult to reduce to levels that are satisfactory for accurate angular measurement. Moreover, the overall length of the telescope would have to be increased and this is undesirable. Another disadvantage of the anastigmatic triplet in the context of EDM is that at the trunnion axis the infra-red signal must be removed from the telescope but the beam-width of the signal is large, so a large beam-splitter prism is needed and the diameter of the trunnion axis must also be large. These difficulties are overcome to some extent by the use of the Cassegrain telescope with catadioptric components described in principle in section 2.1.11. An erect image is produced by a Pechan prism (section 2.1.10.1).

An interesting aspect of kinematic design (section 2.2) applied to the construction of the telescope is described by Moore and Sims, 1980. They point out that to achieve 2″ pointing accuracy with the HP 3820A, the Mangin mirror should be stable to 1″, corresponding to a variation of 0.25 μm across its diameter. Moreover, the mount should not exert any appreciable force on the mirror because a distortion of only 0.06 μm could cause noticeable loss of resolution. Figure 3.3 shows a cross-section of the stainless steel cell containing the mirror.

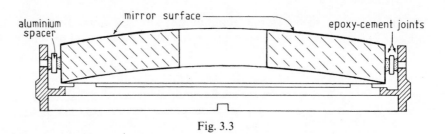

Fig. 3.3

The mirror rests on three pads (the minimum number necessary to constrain it) and is retained by aluminium spacers and epoxy-cement layers. The thicknesses of the spacers and layers were chosen so that the combined thermal expansion of the glass, the aluminium and the epoxy-cement exactly matches that of the steel cell. The cell itself is also semi-kinematically mounted inside the telescope housing.

The magnification of the telescope is 30✕ and the objective aperture is 66 mm, with a field of view of 1.5°.

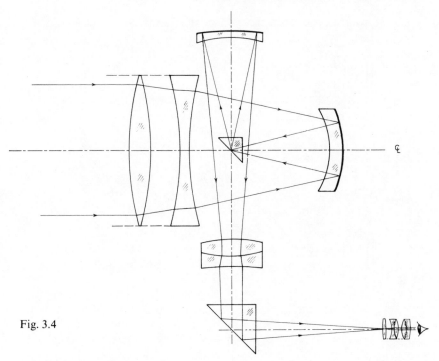

Fig. 3.4

3.1.3 The telescope of the Kern DKM 3 Precision Theodolite

This theodolite is another which makes use of a catadioptric system. It is illustrated in Fig 3.4. In this case, a real intermediate image is formed by the objective doublet and first Mangin mirror. This image serves as an object for the second Mangin mirror and focusing lens. The final real, upright image is formed at the reticule where it is viewed through the eyepiece in the usual way.

This particular design allows for a large aperture (68 mm) and magnification (45×) within a relatively short telescope. This is achieved by taking the rays out of the telescope barrel through the hollow steel trunnion axis and thence to the reticule and eyepiece by reflection in a 45° prism. The additional reflection at the second mirror ensures that the final image is upright. The field of view is just over 1° and the shortest focusing distance is 5 m.

One of the disadvantages of having a short telescope and a small field of view is that it is often difficult to find the target through the telescope. The DKM 3 is provided with a finder telescope to overcome this problem. There is only one eyepiece for the two telescopes, the axes of which are parallel and rotate together about the trunnion axis. A switch next to the eyepiece allows the operator to change from the finder telescope to the main telescope and *vice versa*.

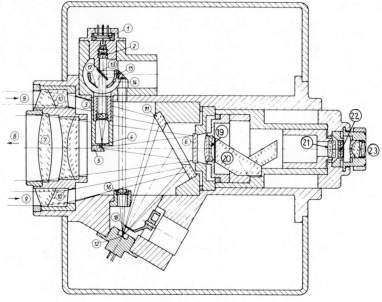

Fig. 3.5

Fig 3.5 Key

1. gallium arsenide diode
2. collimating lenses
3. focusing lenses
4. image of diode and focal plane of telescope objective
5. mirror
6. infra-red/visible light beam-splitter plate
7. telescope objective
8. outgoing infra-red signal
9. incoming infra-red signal
10. annular objective
11. annular mirror
12. avalanche photodiode
13. infra-red beam-splitter plate
14. automatic shutter to divide external and internal (calibration) signals
15. prism for internal calibration path
16. focusing lenses
17. automatic filter
18. automatic filter
19. objective lens
20. Abbé prism
21. focusing lens
22. reticule
23. eyepiece

3.1.4 The telescope of the Wild Tachymat TC1

The TC1 is an electronic tacheometer. The telescope has an objective lens system consisting of two main coaxial components. The inner component is an anastigmatic triplet with an aperture of 34 mm. It is illustrated in Fig 3.5 (7). The outer component (10) is an annular achromatic doublet of aperture 46–59 mm. The source of the infra-red signal for EDM is a GaAs diode (1). The radiation is focused by the lens systems (2) and (3) to focal point (4). This point lies in a focal plane of the objective (7) and glass plate (6) combination. This glass plate has a reflection coating which transmits visible light but reflects infra-red. The infra-red radiation focused to (4) on the surface of prism (5) is reflected from plate (6) and emerges (8) parallel to the optical axis after refraction through the objective (7). After reflection from the target, that part of the infra-red signal which falls on the annular objective (9) is refracted to the annular mirror (11) from which it is reflected to a focal point (12) at an avalanche photodiode (section 2.4.2.1). The beam-splitter (13), shutter (14), prism (15) and lens (16) allow for 4% of the infra-red radiation to be focused at the photodiode (12) to define the internal path length for EDM. Filters (17) and (18) ensure an optimum received signal level for phase comparison.

Incident visible light entering the objective (7) passes through the centre of the annular mirror (11) to the lens (19) and Abbé prism (20). The focusing lens (21) is shown in the position for shortest focusing distance and the image of the target is formed at the reticule (22) where it is viewed through the eyepiece (23). The shortest focusing distance is 2 m and the field of view is 1.8° at infinity focus.

3.2 Axis systems

Nearly all modern theodolites have a cylindrical vertical axis made of hardened steel with a ball-race to carry the weight of the alidade. Location of the axis is by semi-kinematically designed non-loadbearing locations.

3.2.1 The Kern vertical axis system

This is illustrated in Fig 3.6. The weight of the alidade and telescope is supported by a precision ball-race at the rim of the alidade. Centring of the axis is achieved by means of a cylindrical stub which is free of load and located in a sleeve. It can therefore be short and have a small diameter. This arrangement gives a relatively small overall height and increased stability. Haller, 1963 gives the standard error of the

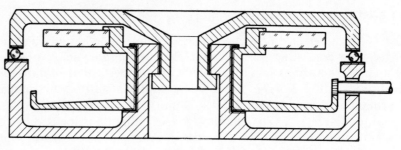

Fig. 3.6

verticality of the axis around the circle as ± 0.35″ for the Kern DKM 2.

3.2.2. The trunnion axis of the Hewlett-Packard 3920A Total Station

This is illustrated in Fig 3.7. In use, the axis is supported by two pads which approximate to the kinematically designed V-block bearing (section 2.2). The third retainer pad prevents the axis from falling out of the bearing when the instrument is packed up and transported. Kerschner, 1980 quotes a value of 1.5″ for the maximum value of the wobble of the trunnion axis in its bearing.

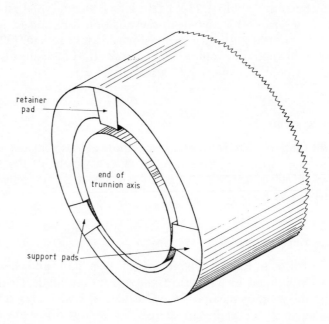

Fig. 3.7

3.2.3 The vertical axis of the Wild T1 Micrometer Theodolite

This is an example of the double centre cylindrical axis which is in widespread use in modern theodolites. It is illustrated in Fig 3.8. The axis is hollow and contains the objective of the optical plummet and the reflecting prism. The precision ball-race (3) is of relatively small diameter and it supports the alidade (2). Constraint of the alidade is also provided by the flange at the lower end of the axis. The horizontal circle (1) is mounted on a sleeve which lies outside the axis.

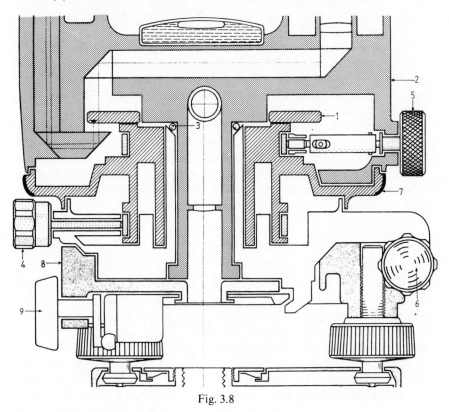

Fig. 3.8

3.3 Movement controls

3.3.1 The horizontal movement controls of the Wild T1 Micrometer Theodolite

These are illustrated in Fig 3.8. The alidade (2) can be clamped to the sleeve which carries the horizontal circle (1) by means of the upper plate clamp (5). The horizontal circle sleeve can be clamped to the fixed axis by the lower plate clamp (4). Each of the clamps has a slow-motion drive associated with it. The lower plate slow-motion drive screw (6) allows the user to rotate the circle relative to the fixed axis. The upper plate slow-motion drive screw cannot be shown in the sectional diagram of Fig 3.8, but it is used to rotate the alidade relative to the horizontal circle. A slow-motion drive is effective only if the respective clamp is on. In normal use during the measurement of horizontal angles, the lower plate clamp and slow-motion drive are not used because the circle must remain stationary. To warn the user against accidental use of the lower plate knobs, the knurling around their circumferences is different from that used for the upper plate knobs. It is therefore possible to sense by touch alone which control knobs are being used.

The double centre system can be used for the measurement of an angle by repetition (when the circle is carried round with the alidade for alternate pointings) and for reiteration, which is the more usual method of measuring an angle using different initial settings of the horizontal circle. The black milled ring (7) around the lower part of the theodolite can be used to rotate the horizontal circle about the axis directly, to obtain an approximate setting. Accurate setting, if needed, can then be made using the lower plate clamp and slow-motion screw.

A clamping lever (9) allows the theodolite to be removed from the tribrach and replaced by another Wild unit, such as a target, which is then located by constrained centring (sections 3.6 and 5.6).

3.3.2 The horizontal movement controls of the Wild T2 Universal Theodolite

The axis system and movement controls are illustrated in Fig 3.9. The horizontal circle (1) on its carrying sleeve rests outside the fixed axis. The weight is supported by a small flange protruding from the casing at the upper part of the sleeve. The alidade (2) is supported by the ball-race (3) and is constrained within the hollow steel axis. A clamp (4) connects the alidade to the fixed axis. When this clamp is on, the slow-motion drive screw (5) can be used to rotate the alidade relative to the fixed axis and to the horizontal circle. A drive screw (6) acts directly on the horizontal circle carrier. Accidental rotation during measurement is prevented by the hinged cover plate (7).

This type of construction allows for measurement by reiteration only as there is no mechanism for carrying the circle round with the alidade. The clamp (9) releases the theodolite from the tribrach.

3.3.3 The movement controls of the Zeiss (Jena) Theo 010A

Instruments in the Zeiss (Jena) Geomat range have clamps and slow-motion screws similar to those illustrated in Fig 3.10 which shows the

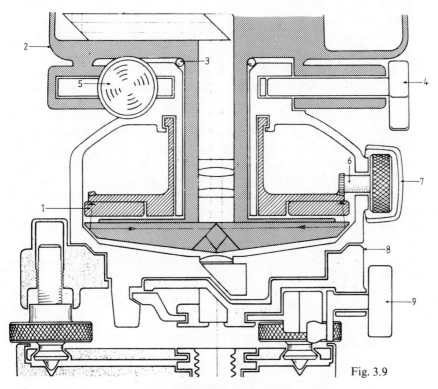

Fig. 3.9

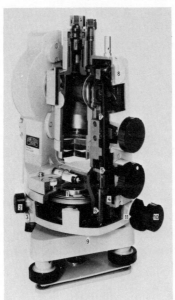

Fig. 3.10

Theo 010A Seconds Theodolite with some cut-away sections. The main features of the movement controls are the 'clothes-peg' arrangement of the clamps and the co-axial slow-motion screws. The horizontal circle (1) can be rotated about its axis by the drive knob (2). Before the drive can be engaged, the safety catch (3) must be swung away to allow knob (2) to be depressed. The alidade clamp is the lower (4) of the pair of levers; the upper lever (5) is the clamp for the telescope. The clamps are both 'on' in the figure. Movements of these clamping levers are transmitted to the respective components by means of the pivoted bars (6) and (7) for the alidade and telescope respectively. Adjustment of the clamping pressure can be made at screws lying behind the covers (8) and (9). This adjustment becomes necessary from time to time because the amplitude of the movement of the clamping levers is not sufficient to compensate for normal wear of the clamping pads. The more common screw clamps, on the other hand, have sufficient amplitude of movement to compensate for this normal wear.

The inner slow-motion screw (10) is for the alidade and the outer (11) is for the telescope.

3.4 Circle reading systems

Examples of each of the four main methods of reading the circles are given. These methods are: direct reading; optical scale reading; optical micrometer reading; and optoelectronic scanning with conversion of data to an equivalent angular reading by a controlling microprocessor. Direct reading is used in the simplest models and the smallest unit of graduation is usually 5' or 10'. Scale reading is used in middle-order theodolites and the smallest graduation is usually 1', but it can sometimes be as small as 20". In higher-order theodolites, optical reading is usually by micrometer with 1" graduations. Optoelectronic scanning is carried out usually with a least count of a few seconds of arc.

3.4.1 Direct reading

Two examples are given. Figure 3.11 shows the appearance of the vertical and horizontal circles seen through the circle-reading microscope of the Kern KOS Construction Theodolite. The horizontal circle reading is 53° 12' by interpolation between the 53° 10' and 53° 15' graduations. The vertical circle reading is 85° 48' which corresponds to a vertical angle of +4° 12'. Alternatively, the upper scale indicates a slope of +7.35%. This direct reading of gradient in the form of a percentage is particularly useful when setting out or measuring gradients on construction sites.

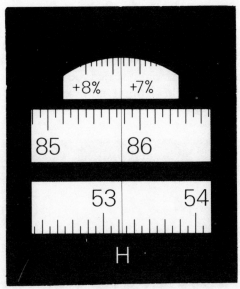

Fig. 3.11

Figure 3.12 illustrates the field of view of the Zeiss (Oberkochen) Th51 Minute-Reading Theodolite. The horizontal circle reading is 70° 04′ (by interpolation between the 70° 00′ and 70° 10′ graduations) and the vertical circle reading is 90° 48′, which indicates a vertical angle of − 00° 48′.

3.4.2 Optical scale reading

The Wild T16 Scale Reading Theodolite is made with four different circle graduations. Circles with the conventional sexagesimal and centesimal units are available and a circle graduated 0–6 400 mil is also produced. In each of these three instruments, the graduations run clockwise around the horizontal circle. In addition, a model is produced with sexagesimal units running anti-clockwise as well as clockwise around the horizontal circle. This can be used for setting out

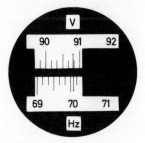

Fig. 3.12

by turning angles either clockwise or anti-clockwise from a given direction without the need to make additional subtractions. The circle readings illustrated in Fig 3.13 are for the centesimal model. Readings are 214.964^g (horizontal) and 94.064^g (vertical, indicating a vertical angle of $+5.936^g$). The horizontal scale is coloured yellow to make the distinction between the horizontal and vertical circle scales more clear.

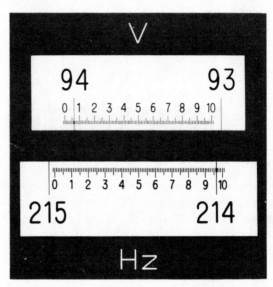

Fig. 3.13

The Zeiss (Jena) Theo 020A sexagesimal circle readings are illustrated in Fig 3.14. The main interval of the scale is 1', but each of these is divided into $3 \times 20''$ intervals, with the 20'' and 40'' graduations displaced laterally to avoid a confusion of graduation lines. The readings are 262° 08' 20'' (horizontal) and 138° 07' 20'' (vertical, indicating a vertical angle of −48° 07' 20'').

In each case, with scale-reading theodolites, the main figure (degrees or grades) is the figure corresponding to that main scale graduation which intersects the scale. There is an ambiguity when the reading is an integral number of degrees, when the higher value of the main scale graduations intersects the scale at the zero mark and the lower value graduation intersects the scale at the higher (60' or 100^g) mark. When the reading is a minute or two either above or below an integral number of degrees (or grades) the scale graduations extended just beyond the zero and 60' (or 100^g) marks allow the user to make a reading at either end of the scale, and care should be taken to ensure that the circle is not misread by a whole degree (or grade).

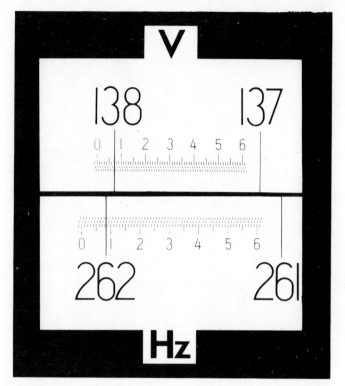

Fig. 3.14

3.4.3 Optical micrometer reading

The Wild T2 Universal Theodolite makes use of coincidence reading to eliminate circle eccentricity errors. In this section, the optical micrometer reading system for the horizontal circle of the T2 is described.

In Fig 3.15, VV′ represents the vertical axis and HH′ represents the trunnion axis. A cylindrical bundle of rays falls on the mirror M and passes normally into prism (1), emerging vertically and symmetrical about the vertical axis VV′. After passing through a converging lens (2) the bundle meets the compound prism (3) and (3′) which splits the beam into two parts, each bundle being brought to a focus at the graduated circle CC′ onto diametrically opposite graduations GD and G′D′ (assuming the two graduations happen to fall in the plane of the diagram).

The central ray of the bundle converging at A on GD is reflected from the silvered circle into prism (3) again and passes through the objective (4), is refracted laterally by the rhomboid prism (5) and falls on a parallel-sided glass plate (6). The central ray from A′ on G′D′

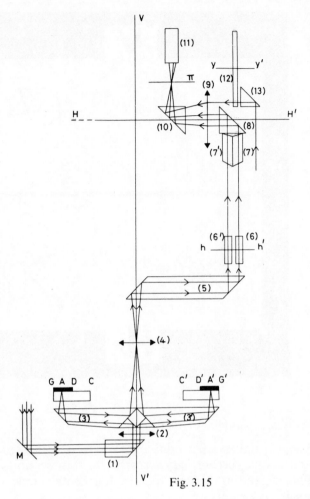

Fig. 3.15

follows a symmetrical path and falls on plate (6'). These two plates are part of the optical micrometer and can be rotated about a horizontal axis hh'.

After passing through (6) and (6') the two bundles of rays pass through prisms (7) and (7') (i.e. the separator system) then through prism (8), converging lens (9), prism (10) and form real images of A and A' respectively in plane π which are examined through the circle reading microscope (11).

3.4.3.1   The separator system

Consider first, prism (3) and images of the graduation GD (Fig 3.16). Prism (3) acts as two plane mirrors inclined at 45° and as a result of

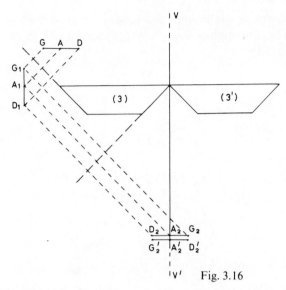

Fig. 3.16

reflections at its sloping faces, an image is formed first at G_1D_1 and secondly at G_2D_2.

If the thickness (t) and refractive (μ) of the glass of prism (3) are taken into account, G_2D_2 will simply be shifted along VV′ by an amount equal to $t(1 - 1/\mu)$. Similarly an image $G'_2D'_2$ of G′D′ will be superimposed on the image G_2D_2.

Suppose now that the graduations GD and G′D′ are limited to segments AD and A′D′ respectively. Their images will, ideally, coincide at A_2 and A'_2 (Fig 3.17a). However, it is impossible to limit the length of the circle graduations to this accuracy, so in practice A_2 and A'_2 will either overlap (Fig 3.17b) or be separated (Fig 3.17c).

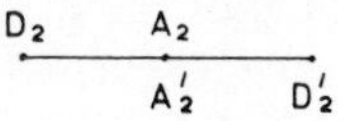

Fig. 3.17a

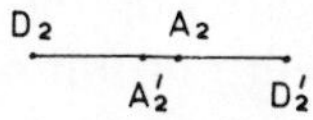

Fig. 3.17b

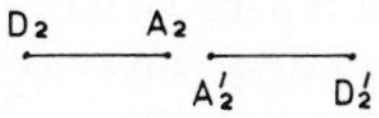

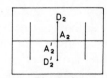

Fig. 3.17c

The resultant final image at π (viewed through the microscope (11)) will be as shown in Fig 3.18a if A_2 and A'_2 overlap or as shown in Fig 3.18b if A_2 and A'_2 do not meet. The ideal image is shown in Fig 3.18c. In order to achieve this final image, the separator system (7) and (7′) is

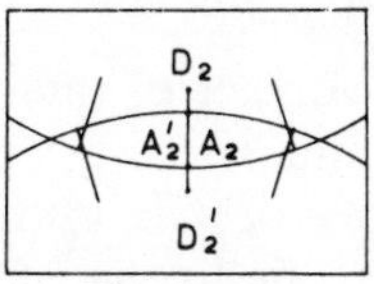

Fig. 3.18a

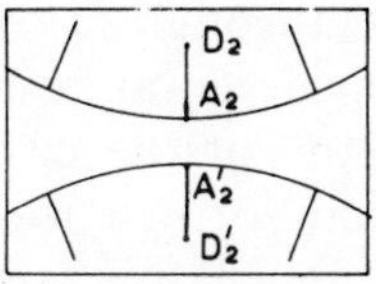

Fig. 3.18b

Fig. 3.18c

employed. Suppose the images are separated as shown in Fig 3.18c. After refraction through prism (5), the situation will in effect be that shown in Fig 3.19.

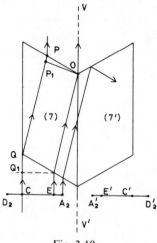

Fig. 3.19

Only segment CE of the image A_2D_2 will be passed by the prisms (7) and (7'); the vertical ray from C will just meet prism (7) and will emerge as shown at P. The vertical ray from A_2 will be totally internally reflected at the upper face of prism (7') and will be sent out of the optical train. A similar segment (C'E') of image $A'_2D'_2$ will be transmitted and the important feature is that the rays from E and E' will coincide along the axis of the prism after refraction. A similar result will be achieved if A_2D_2 and $A'_2D'_2$ overlap.

If the position and focal length of lens (4) are suitably chosen, the images of E and E' will fall at O on the outer face of the prism block. The image of C however will not fall on this face at P (owing to the optical path differences) but at P_1 where $\mu(PP_1) = QQ_1$. Thus the face OP has to be shaped along OP_1 and not along OP. The image of CE is formed at the face OP_1, and similarly for the image of C'E'. In practice this will not introduce any problems in focusing the microscope on the inclined planes since their inclination is about 65° to VV' (instead of 90°) and the images of CE and C'E' are only about 0.1 mm long. Each of the faces carries an index (I and I') engraved at right angles to the junction of (7) and (7') and through O.

These indices can be considered as being the images of imaginary indices J and J' in the plane of the circle, I and I' being formed by that same system which formed the images of CE and C'E'. These imaginary indices J and J' are assumed to lie in the plane of Fig 3.15.

3.4.3.2 The optical micrometer

It has been assumed so far that a pair of diametrically opposite graduations falls in the plane of Fig 3.15. In general, this will not be so and the imaginary indices J and J' will lie between circle graduations.

In Fig 3.20 the circle reading at J is $(36°\ 20' + \epsilon)$ and at J' it is $(216°\ 20' + \epsilon')$, ϵ' being equal to ϵ if there is no eccentricity which is assumed for the time being.

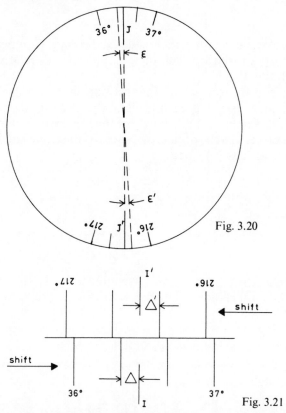

Fig. 3.20

Fig. 3.21

The images of the two arcs at the exit faces of the separator system will be as shown in Fig 3.21 where I and I' are the real index lines. It is necessary to measure the linear distances $\triangle$ and $\triangle'$ corresponding to the angular differences ϵ and ϵ' respectively which are for the time being assumed equal. To measure $\triangle (= \triangle')$ the parallel plates (7) and (7') are rotated about their axis hh' by equal amounts but in opposite directions. Each plate displaces the image formed by the rays passing through it, by equal amounts but in opposite directions (see section 2.4.1.3). Thus plate (7) displaces the image of the circle near J relative to I as shown in Fig 3.21. Plate (7') has the opposite effect on the image

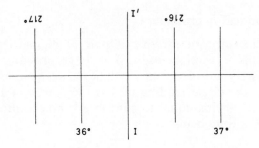

Fig. 3.22

of the circle near J′. Thus the image can be made to appear as shown in Fig 3.22 by translating each image a distance $\triangle$ relative to the fixed indices I and I′.

It is necessary to measure the translation $\triangle$ in order to deduce the angular value ϵ. If R is the radius of the horizontal circle and R′ the radius of its image at the exit face of the separator system and if D and D′ respectively are the optical distances of the circle and its image from the objective lens (4), then

$$\frac{R'}{R} = \frac{D'}{D} \text{ and } \triangle \simeq R'\epsilon$$

Therefore

$$\epsilon \simeq \triangle \cdot \frac{1}{R} \cdot \frac{D}{D'}$$

It has already been shown (section 2.4.1.3) that for refraction through a parallel-sided glass plate, thickness t, refractive index μ and rotation i,

$$\triangle \simeq ti\left(1 - \frac{1}{\mu}\right)$$

Thus

$$\epsilon \simeq \frac{ti}{R}\left(1 - \frac{1}{\mu}\right)\frac{D}{D'},$$

so that ϵ is approximately proportional to i, the angular rotation of the parallel plate.

The maximum value of the rotation ϵ necessary to bring a graduation into coincidence with an index is one half the circle graduation interval, i.e. 10′. Thus the maximum value of i is about 12°. If the value of ϵ is to be read to the nearest second, then it can be seen that it is very difficult to graduate a 12° arc of the micrometer control to represent a range of 10′ in intervals of 1″ without having an inconveniently large mechanical system. There is therefore a micrometer circle with the 1″ graduations from 0′ to 10′ around an arc of $\theta°$ of its circumference connected mechanically to the micrometer control by a gearing of $12°/\theta°$. This micrometer scale is shown as plate (12) in Fig 3.15, rotatable about the horizontal axis yy′ as the parallel plates (6) and (6′) are rotated about their common axis hh′. This micrometer scale is numbered in minutes and seconds every 10″, each 10 second interval being subdivided into 10 one second intervals. Estimation is to 0.5″.

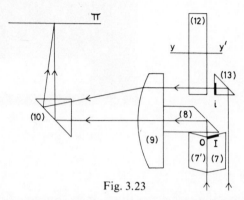

Fig. 3.23

Fig 3.23 illustrates the path of the rays through the micrometer. Prism (13) carries the micrometer index i. Plane π contains the images of both the main scale and the micrometer scale. A mask placed at the objective lens (9) excludes unwanted parts of the images so that the final image at π which is examined by the microscope is as shown in Fig 3.24.

Figure 3.25 shows a diagram of the micrometer and separator assembly. Figure 3.26 illustrates the field of view through the circle-reading microscope of a more recent version of the T2 where the coincidence reading is obtained in a similar way to that described above, but the micrometer and main scale graduation values have been

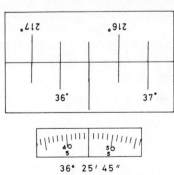

36° 25′ 45″

Fig. 3.24

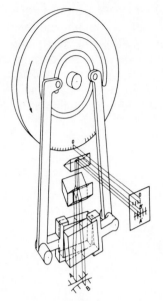

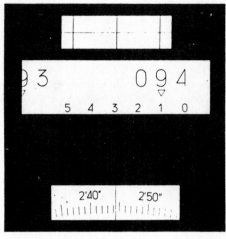

Fig. 3.26

Fig. 3.25

redesigned so that the circle reading can be obtained directly to 10″ without the need to count the 10′ intervals as is the case in the older version. The circle reading illustrated is 94° 12′ 44″, to the nearest 1″.

It has been assumed so far that the line joining the imaginary indices J and J′ is a diameter of the horizontal circle and that the centre of rotation of JJ′ coincides with the centre of the circle graduations. In practice, this will not always be so. The following description indicates how the elimination of eccentricity errors dealt with in general terms in section 5.3 is carried out in the Wild T2. Suppose (Fig 3.27) that the imaginary indices J and J′ are misplaced relative to the centre of the circle and lie instead at J_1 and J'_1. Then J_1 and J'_1 are eccentric to the centre of the graduations by an angle e.

The image at the exit face of the separator system is then as shown in Fig 3.28.

SS′ is the axis of symmetry between the graduations corresponding to 36° 20′ and 216° 20′. The linear parts $\triangle_1$ and $\triangle'_1$ are unequal. If R′ is the radius of the circle, then

$$\triangle_1 = R' \epsilon_1 \text{ and } \triangle'_1 = R'\epsilon'_1,$$

$$d = R' e \text{ and } \triangle = R'\epsilon.$$

But,

$$\frac{\epsilon_1 + \epsilon'_1}{2} = \epsilon$$

Therefore, substituting for ϵ_1, ϵ'_1 and ϵ

$$\frac{\Delta}{R'} = \frac{1}{2}\left(\frac{\Delta_1}{R'} + \frac{\Delta'_1}{R'}\right)$$

Therefore,

$$\Delta = \frac{\Delta_1 + \Delta'_1}{2}$$

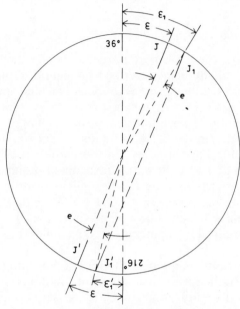

Fig. 3.27

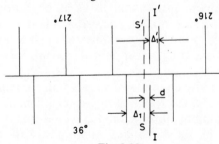

Fig. 3.28

Thus, the semi-sum of the linear parts Δ_1 and Δ'_1 is equal to Δ, which is the value without eccentricity.

When the parallel plates are rotated, the upper image in Fig 3.28 is translated through a distance $(\Delta'_1 + d)$ and the lower image is translated through a distance $(\Delta_1 - d)$. Coincidence between the 36° 20′ and the 216° 20′ graduations is established on the axis of symmetry SS′, the distance translated being Δ for each image. There is therefore no need to have two indices I and I′, since in practice coincidence is established between diametrically opposite graduations. In fact, only one index (I) is present, this being used solely to indicate the approximate centre of the field of view and hence the approximate reading.

Types of coincidence settings and their relative accuracies are described by Smith, 1970.

3.4.4 Optoelectronic systems

In the last few years, there has been a significant change in theodolite design which has already been described in general terms in section 2.4.2. In this present section, some of the ways in which optoelectronic components are used in instruments in current production are described. The intention here is not to describe each or even one of the electronic instruments in detail, but to describe one or two particular features of some of the instruments. The derivation of circle readings in a digital electronic form allows the instrument designer to extend the capabilities of the theodolite far beyond its traditional use as a device which allows the observer to read and write down numerical values. The examples which follow in sections 3.4.4.1 etc. illustrate some of the current developments of electronic theodolites and tacheometers.

There is no generally accepted method for coding and scanning the circles. Some manufacturers use different methods for different instruments. This diversity of approach is likely to continue, but as electronic components become less expensive relative to optical and mechanical components and as the users become more familiar with the automatic transfer and processing of data and with the opportunities these offer, it is possible that just a few methods will become reliable, accurate, economical to produce and accepted generally.

With each theodolite having optical methods for reading the circles, the operator controls the circle-reading mechanism and is responsible for manual recording of observed data, even if this recording is done by entering data at the keyboard of a solid-state memory device instead of in a field-book. In working with an electronic theodolite or tacheometer, however, at least the direct control of the reading device is taken away from the operator and is carried out by the controlling microprocessor which works according to instructions in its memory, prompted by the operator. Data are usually displayed so that the operator can record them manually in a field-book in the conventional way if required, or they can be stored automatically, again under the control of a microprocessor prompted by the operator.

Each of the instruments described in sections 3.4.4.1 etc. can be used to measure distances as well as angles. In some, the EDM components are integral with those used for measuring angles and such an instrument is an *electronic tacheometer*. In others, an EDM module can be added to an *electronic theodolite* and usually a third module is available in the form of a solid-state memory device with its own keyboard and display. The memory device usually has its own microprocessor which is separate from the controlling microprocessor that controls the measurements. The data storage device and its microprocessor can often be used to carry out not only data storage but also computations of traverses, resections etc. at the instrument. In such a case, the module is referred to as a *field computer*.

Detailed discussion of EDM is covered by Burnside, 1971 in a companion volume and is not attempted here. In describing particular features of an instrument, however, it is thought desirable to include a general description of the system of which the particular features described form a part and EDM is considered in this context only.

3.4.4.1 The field computer of the Zeiss (Oberkochen) Elta 2 Recording Computer Tacheometer

The Elta 2 is an electronic tacheometer. It derives from the RegElta 14 which was first produced in 1968. The Elta 2 is illustrated in Fig 3.29 and is the first 'second generation' electronic tacheometer.

Fig. 3.29

In its basic form it consists of an electronic theodolite with integral EDM by amplitude modulated infra-red radiation. The telescope is used as the sighting device for angular measurements and also for the transmission and reception of the returned EDM signal. It has an objective aperture of 60 mm and a magnification of 30×. The 98 mm diameter circles are coded identically at an interval of 0.5^g which is effectively halved by the optoelectronic scanning of diametrically opposite sectors. The coding is absolute, not incremental. The Elta 2 uses a rotating parallel-plate micrometer to maintain automatically the diametrically opposite images of the circle coding marks. The micrometer turns continuously during measurement and the circle code and current micrometer value are combined in a full circle reading. This automatic system is therefore very similar to the conventional optical micrometer system described in section 2.4.1.3, except of course that the coincidence is detected and maintained automatically in the case of the Elta 2 with the result that 'readings' are made more quickly and without personal bias. The same micrometer is used for both circles and the light paths associated with the projection of the vertical circle pass through a pendulum compensator which corrects the reading for residual tilts of the vertical axis in the plane of the telescope. The micrometer is coded to give a resolution of 1/1 250 of the 0.25^g effective main graduation interval, so the overall resolution is 2.0^{cc}, or about 0.6″, and the manufacturers quote the same figure

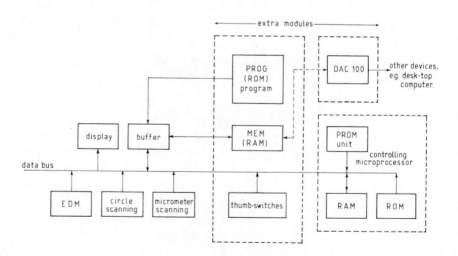

Fig. 3.30

(±0.2 mgon) as the standard error of the mean of a face-left and a face-right direction. The EDM range is up to 4 km (with nine prisms) and is said to be to a precision of ±10 mm.

The controlling microprocessor can be used to carry out the usual functions (measurement, reduction and display of data) including zeroing the horizontal circle, reversing the direction of the horizontal circle increments, correcting the EDM for slope, earth curvature and refraction and computing the horizontal distance and height difference for a measured line.

The Elta 2 electronic tacheometer described above can be converted to a recording computer tacheometer by the addition of the devices PROG and MEM, which are respectively a series of programs in ROM modular form and a solid-state memory. Figure 3.30 illustrates schematically the relationship between the controlling microprocessor and the additional components corresponding to PROG and MEM. A 14-digit set of switches is also added with PROG and MEM to the basic Elta 2 so that the operator can give instructions for the execution of the programs and storage of data.

The addition of the PROG1 memory module allows the operator to carry out the following additional computations at the instrument.

1. Calculation of the heights of the target stations.
2. Reduction of distances to the projection.
3. Determination of and allowance for collimation and index errors in the circle readings.
4. Calculation of eastings and northings of the target stations.
5. Resection of the instrument position from measured directions to three control points.
6. Resection of the instrument position from measured directions and distances to two control points.
7. Resection of the instrument position by a least-squares determination of the parameters of a Helmert transformation, using measured directions and distances to three or four control points.
8. An orientation check on the horizontal circle after 5, 6, or 7.
9. Following 5, or 6, or 7, and 8, the determination of the eastings, northings and heights of target stations.
10. Computation of the co-ordinates of successive stations in a traverse.
11. Computation of setting-out data to locate a point with given co-ordinates from a trial point, in either rectangular or polar co-ordinates.

The recording unit MEM can be used to record automatically the data computed by the execution of the programs in PROG, or it can be used to record data only when the operator wants to. Data are recorded in blocks. Each block consists of three address digits, fourteen coding digits and three measured or computed values – horizontal direction, difference in elevation and horizontal distance, for example. The MEM device has its own battery which is recharged automatically whilst the Elta 2 is working. This ensures that the data are retained in MEM for a few days after the device has been removed from the instrument. The standard MEM device has three memory units and each can store 200 data blocks.

Data from the memories of MEM can be transferred to another device for further processing and display via the DAC-100 data converter which has standard sockets for TTY (teletype) and RS 232/V24 connections. This data converter also allows the user to store data in MEM for later recall on site. In this way, the co-ordinates of control points and points to be set out can be stored at specific locations in the RAM unit and then recalled and used in conjunction with computations 5 and 11 above, for example, to derive setting-out data.

It should be clear that the Zeiss Elta 2 used with the additional modules PROG and MEM becomes not just a tacheometer, but a site computer as well. In order to take full advantage of the possibilities offered by the programs, the operator will have to learn to combine measurements and computations, two aspects of surveying which have often through necessity been executed at different times. The possibility of combining them on site should result in improved efficiency, but it is not easily achieved in practice without the co-operation of an operator with an understanding of the logical processes of computers.

3.4.4.2 The circle scanning system of the Wild Tachymat TC1

The Wild Tachymat TC1 is an integrated electronic tacheometer made in conjunction with SERCEL and first produced in 1977. Its general appearance is illustrated in Fig 3.31. Distance measurement is by infrared radiation transmitted and received through the telescope (see section 3.1.4 and Fig 3.5). The microprocessor controls the measurements of angles and distances and also reduces these measurements to allow for scale factor, additive constants, atmospheric refraction, earth curvature and slope. Horizontal distances, height differences, heights above datum and plane rectangular co-ordinates can also be computed and displayed by the microprocessor. Commands are entered by the operator at a keyboard. This keyboard is on the alidade, so it is duplicated at the opposite side. Commands can then be entered easily in either the face-right or the face-left

Fig. 3.31

position. Standard blocks of either measurement data or code data can be recorded on a magnetic tape cassette housed in a unit which is located across the tops of the standards. Each cassette can hold about 1 800 blocks of data. An interface having standard TTY (teletype) and RS 232/V24 connections can be supplied so that the recorded data can be transferred to another device for further processing. Baud rates from 110 to 9 600 can be used for this transfer of data.

The principle of the method of scanning the circles is illustrated in Fig 3.32 where an electroluminescent diode source of infra-red radiation (1) is shown schematically. Radiation from this source is collimated by the lens (2) and falls on a phase grating – the analyzer grating (3). Only the two first-order maxima of the diffraction are shown, each making an angle ϕ with the normal to the grating. Another grating (4) with spacing equal to half that of the analyzer grating (3) produces first-order maxima $+-$ and $++$ from the incident $+$ beam and first-order maxima $-+$ and $--$ from the incident $-$ beam. These first-order maxima have diffraction angles of 2ϕ because of the half spacing of (4) compared with (3). The $++$ beam and the $--$ are diffracted out of the system, but the $+-$ and the $-+$ beams are incident upon another analyzer grating (3′) similar to (3) where again, only the

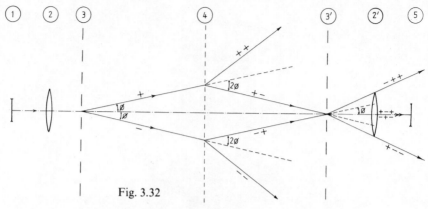

Fig. 3.32

first-order diffraction maxima are shown. Incident beam $+-$ gives diffraction maxima $+-+$ (along the axis of the system) and $+--$. Incident beam $-+$ gives diffraction maxima $-+-$ (along the axis) and $-++$. The two beams which coincide along the axis are focused by lens (2′) onto the photodiode (5). These two beams can interfere. If the phase grating (4) is moved laterally with respect to the rest of the system, the phases of the $+-+$ and $-+-$ beams will be changed in different ways and as a result of their interference the photodiode will detect sudden changes in the intensity of the combined signal. The

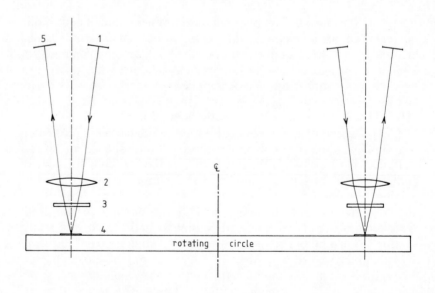

Fig. 3.33

effect at the photodiode is an intensity modulation of the light source, with period half that of the grating (4).

In practice, the phase grating (4) is a radial reflective phase grating on the circle (horizontal or vertical) and the analyzer gratings (3) and (3′) in Fig 3.32 are one and the same radial grating (3) in Fig 3.33. Diametrically opposite sectors of the circle are scanned by identical systems. One important difference between the principle of the method described above and its practical application arises from the radial nature of the gratings. The beam from (1) in Fig 3.33 passes through the radial grating (3) nearer to the centre of the circle than does the diffracted beam. This means that the slit width for the incident beam is slightly less than it is for the return, diffracted beam. Therefore, the interfering beams $+-+$ and $-+-$ do not give sudden changes from maximum to minimum intensity as the circle is rotated, but instead produced a fringe pattern similar to Moiré fringes. An array of four photodiodes at (5) is used to detect changes in the fringe pattern in a way similar to that described in section 3.4.4.3. This method gives an interpolation interval of 0.125 of the fringe period.

The circle has a diameter of 80 mm. The grating (4) consists of 12 500 intervals and at the circumference of the circle, this corresponds to a width of about 20 μm. The reflecting layer is only about 0.25 μm thick. The angular equivalent of a 20 μm interval at the circumference is about 1′ 44″ (or 3.2^c). Interpolation to 1/32 of this value is possible because: firstly, the half spacing of the circle grating relative to the analyzer grating gives an interpolation factor of 1/2; secondly, scanning of diametrically opposite sectors of the circle effectively doubles the movement and gives another interpolation factor of 1/2; and thirdly, the array of photodiodes and the method of analysing their outputs which had already been referred to gives an interpolation factor of 1/8. Thus the scanning is carried out to 1/32 of the interval on the circle, which is equivalent to about 3″ (or 10^{cc}).

The manufacturers claim that the standard error of a horizontal direction derived from the mean of a face-left and a face-right measurement is $\pm$2″. Also, the standard error of a vertical angle from the mean of face-left and face-right measurements is said to be $\pm$3″, the slight increase being the contribution to the circle reading error of the automatic index which defines the horizontal in normal use.

Both circles are coded incrementally. Upon switching on, the horizontal circle is set to zero automatically, but it can be reset to zero at any time by keying-in the appropriate instructions. Positive counting in an anticlockwise direction is also possible. The vertical circle must be referenced (section 2.8) and this can be done in one of three ways. Normally, reference is made to the pendulum which is also used as the automatic index. The procedure is to set up and bisect a target with a horizontal hair, first on face-right and secondly on face-left. The appropriate instructions entered at the keyboard ensure that the mean circle position is taken as the reference direction and subsequent vertical circle readings at that set-up are referred to that mean direction. This method cannot be employed when the instrument is being used in high winds or on a floating platform. In such cases, referencing is made internally to the vertical axis according to the procedures in the handbook. The third method of referencing the circle is to make use of the bubble on the telescope; When this is central, the appropriate instructions at the keyboard set the reference direction to coincide with the telescope axis. It is necessary to ensure that the line of collimation is parallel to the principal tangent of the bubble. Visible warning that referencing has not been carried out is given at the display. After referencing, the scanning of the vertical circle takes place through a pendulum automatic index, so that the effects of small inclinations of the vertical axis are removed from the vertical circle reading.

The maximum speed at which the alidade or telescope can be rotated about their axes without loss of count is two revolutions per second.

Fig. 3.34

3.4.4.3 The circle scanning system of the Hewlett-Packard 3820A Electronic Total Station

This instrument is an integrated electronic tacheometer and is illustrated in Fig 3.34. It was first introduced in 1977. The microprocessor controls and monitors the measurements under commands entered through the keyboard. It also carries out simple transformations of the data to give, for example, the horizontal component of a measured line corrected for slope, atmospheric refraction and earth curvature. Data are displayed by LEDs. It is possible to record data automatically on a small solid-state memory device with its own keyboard and LED display. This is the Hewlett-Packard 3851A Hand-Held Data Collector. Alternatively, the instrument can be connected directly to a minicomputer, such as the Hewlett-Packard 9825A. The output of the HP 3820A consists of 56-bit blocks made up of 14 BCD (binary-coded decimal) digits and the Data Collector can store up to 4 000 such blocks, depending on the storage option chosen. No direct editing of the measured data is possible except when the instrument is interfaced to a computer. This

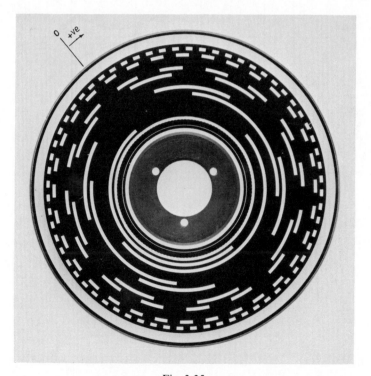

Fig. 3.35

is more likely to be done in a laboratory or workshop than in the field or on a construction site.

There are three different codes on the circles. The vertical and horizontal circles are scanned in the same way. A circle reading is derived by combining the outputs from each of the three systems. Figure 3.35 illustrates the circle and the three codes. The outer code is a 4 096 radial slit track and is too fine to appear as discrete marks in the figure. This track is scanned over diametrically opposite sectors. The central coding pattern is an 8-bit Gray code (section 2.4.2.3) and the innermost coding pattern is a 128-period sinusoidal track.

The 8-bit Gray code effectively divides the circle into $2^8 = 256$ sectors, each about 1.4° (or 1.6^g) wide. These sectors are interpolated

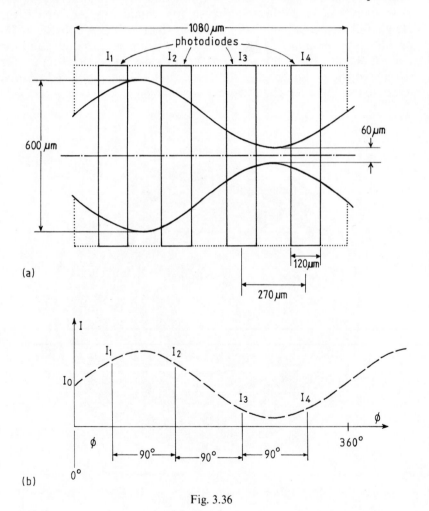

Fig. 3.36

to 1 part in 1 000 by the following method. Four photodiodes (Fig 3.36a and b) constitute an array of sensors of the signal from the sinusoidal track. One period of this track has a length of 1 080 μm at a circle diameter of 44 mm and maximum and minimum amplitudes of 300 μm and 30μm respectively. Each diode has an effective aperture of 120 μm and the separation between the centre of each is 270 μm, that is, one quarter of the period of the sinusoidal track. The photocurrent generated by each photodiode is proportional to the area illuminated. On the assumption that the aperture of each photodiode is narrow (Fig 3.36b) the four output currents are

$$I_1 = I_0 + I \sin\phi$$
$$I_2 = I_0 + I \sin(\phi + \pi/2) = I_0 + I \cos\phi$$
$$I_3 = I_0 + I \sin(\phi + \pi) = I_0 - I \sin\phi$$
and
$$I_4 = I_0 + I \sin(\phi + 3\pi/2) = I_0 - I \cos\phi$$

The differences between alternate output currents are therefore

$$I_1 - I_3 = 2I \sin\phi$$
and
$$I_2 - I_4 = 2I \cos\phi$$

But the incident light is amplitude modulated as a function of time, so that

$$I_1 - I_3 = 2I \sin\phi \sin\omega t$$
and
$$I_2 - I_4 = 2I \cos\phi \sin\omega t$$

The signal $(I_1 - I_3)$ is shifted in phase by $\pi/2$ with respect to $(I_2 - I_4)$ and then the two signals are added to give a total signal I_t:

$$\begin{aligned} I_t &= 2I \sin\phi \sin(\omega t + \pi/2) + 2I \cos\phi \sin\omega t \\ &= 2I (\sin\phi \cos\omega t + \cos\phi \cos\omega t) \\ &= 2I \sin(\phi + \omega t) \end{aligned}$$

This signal is compared with the modulation signal $I \sin\omega t$ to obtain the phase difference, ϕ. This phase discrimination is digital and accurate to 1 part in 1 000. It gives 1 part in 1 000 of the equivalent angular value of one period of the 128-period sinusoidal track, corresponding to about 10″, or 31^{cc}.

It is also possible similarly to discriminate to 1 part in 1 000 of the 4 096 radial slit track, corresponding to 0.3″, or 1^{cc}. The difference is that the sinusoidal track is on the sensors and the bar pattern is on the circle. The diametrically opposite scanning of the radial slit track is used to detect circle eccentricity errors and the output from the sinusoidal track scanning is corrected accordingly.

Figure 3.37 illustrates the optoelectronic couplers used for the scanning. The Gray code provides the coarse reading, so the coding system as a whole is absolute and not incremental, the circle zero

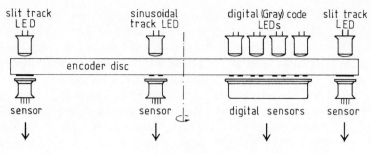

Fig. 3.37

corresponding to the Gray code 00000000 indicated by the radial line in Fig 3.35.

The vertical circle is coded and read in the same way. A tilt sensor (section 3.7.2) is used to determine the inclination of the vertical axis to the vertical. This measured inclination is used by the controlling microprocessor to apply corrections for dislevelment to the horizontal

Fig. 3.38

and vertical circle readings. Referencing of the vertical circle is by its zero graduation which is set at the factory to coincide with the vertical axis. Because tilts of the vertical axis are measured and appropriate corrections applied to the circle readings automatically, it is not necessary to level this instrument as accurately as is usual with other theodolites. There is no tubular plate level, but only a small spherical level on the alidade which allows the operator to set up with the vertical axis within the 2.5′ operating range of the tilt sensor.

The manufacturers claim a root-mean-square error of an angle, deduced from face-left and face-right readings, of $\pm2''$ (or $\pm6^{cc}$) for horizontal angles and $\pm4''$ (or $\pm12^{cc}$) for vertical angles. These errors correspond to the root-mean-square error of ±10 mm at a range of 1 km for the EDM component of the tacheometer.

3.4.4.4 The circle scanning system of the Kern Electronic Theodolite E2

The Kern Electronic Theodolite is illustrated in Fig 3.38 and is the main component of a modular design. The other two modules are the DM 502 EDM device and a semi-conductor storage device (the R32 or R48).

It was first produced in 1981, following the production of the E1 in 1977. It has the same telescope, trunnion axis and automatic index as the Kern DKM 2A optical micrometer theodolite. The DM 502 can be used on either the optical or the electronic theodolite. In the latter case, the slope distance is computed within the DM 502 and transferred to the electronic theodolite where it is transformed and displayed in accordance with the instructions in the ROM of the controlling microprocessor of the electronic theodolite.

A storage device can be attached to one of the standards of the theodolite. This device has its own keyboard and LED display which allow the operator to instruct the microprocessor to carry out certain storage and display operations. No editing of the data at the site is possible. There is no fixed size of a data block for storage, so the operator can store only those data that are needed. This gives a more efficient use of the storage capacity of the memories. On average, data associated with about 600 observations can be stored.

The 70 mm diameter circle is divided radially into 20 000 sectors, each corresponding to an angular value of 2^c, or about 1′ 05″. The images of diametrically opposite regions are superimposed, one image being magnified by a factor of 1.01. The resultant pattern is illustrated in Fig 3.39 and shows Moiré fringes which have an approximately sinusoidal intensity distribution. Four photodiodes are arranged with respect to the sinusoidal pattern in a manner similar to that described

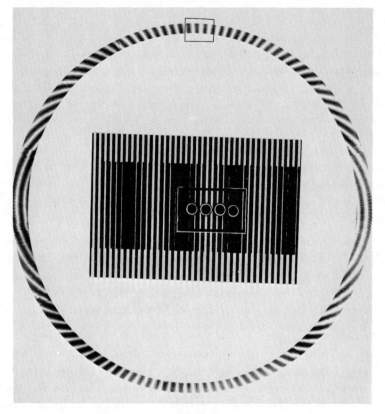

Fig. 3.39

in the foregoing section and covering three-quarters of the period of the fringes. The outputs from these four photodiodes are initially subtracted as described by the equations in section 3.4.4.3 to obtain the differences (putting $I = 1$)

$$I_1 - I_3 = 2\sin\phi$$
and $$I_2 - I_4 = 2\cos\phi$$

The two outputs characterised by these two equations are both treated in two different ways to provide firstly a coarse count of the rotation of the array of photodiodes relative to the circle and secondly an interpolation within a period of the Moiré fringe pattern to give a 'fine reading'.

To detect rotation and measure the coarse count, each signal $(I_1 - I_3)$ and $(I_2 - I_4)$ is used as input to a *Schmitt trigger*. This is a solid-state circuit which converts a relatively slowly varying input signal to an output signal which changes very rapidly from one level to another when the input reaches a certain value (Fig 3.40). The two

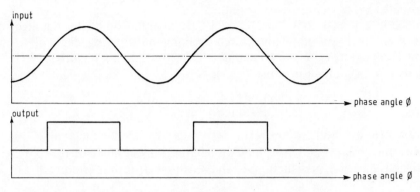

Fig. 3.40

square-wave outputs from the Schmitt triggers are separated in phase by $\pi/2$ so when added, they give four discrete steps for counting within a Moiré fringe period (Fig 3.41). Each step corresponds to one-quarter of an angular division, i.e. 50^{cc} or about 16″. The different senses of rotation of the diametrically opposite sectors of the circle give a further division of the counting interval by a factor of 2, leading to a least count of 25^{cc}, or about 8″. However, only a coarse reading is required,

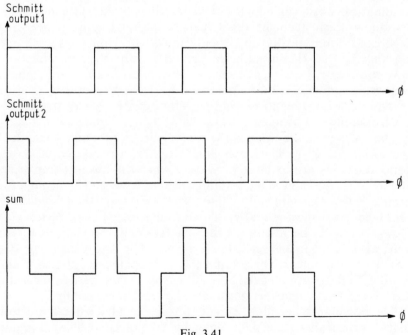

Fig. 3.41

so only every fourth trigger step is significant, the corresponding angular value being 1^c, or about $32''$.

To interpolate to 1^{cc}, or $0.3''$, a 1% discrimination of this counting interval is required. This is achieved by multiplying each of the difference outputs $(I_1 - I_3)$ and $(I_2 - I_4)$ from the diodes with a 1 kHz harmonic signal to produce the signals

$$(I_1 - I_3)' = 2\sin \phi \cos \omega t$$
and $\quad (I_2 - I_4)' = 2\cos \phi \sin \omega t$

When these are added, the resultant is $2\cos(\phi - \omega t)$ which is compared with the $\cos\omega t$ signal in a phase discrimination counter controlled by a 100 kHz oscillator which gives the required discrimination to 1^{cc}, or $0.3''$.

It is necessary to relate these coarse and fine readings to produce an unambiguous registration of the circle position. The coarse reading takes place during rotation and the interpolation during the time that the circle and alidade are stationary. Although the phase angle derived from the phase discrimination is fixed in relation to the Moiré pattern, the absolute position of the zero phase angle within the fringe is not known. The criterion for relating the coarse and fine readings is that the phase reading (in units of $1/100$ of 2π) must change from 99 to 0 when the coarse reading registers a trigger pulse. This location of the zero of the phase reading with respect to a trigger pulse is carried out automatically by a small logic circuit described by Aeschlimann, 1979. As soon as the instrument is switched on, the horizontal circle reading is stored automatically and then subtracted from all subsequent readings so that the initial pointing is effectively the zero direction. Any value can be assigned to the horizontal circle reading by a procedure similar to that used in a double-centre optical reading theodolite with repetition clamp, although the re-registration is electronic instead of mechanical.

The coding and scanning of the vertical circle are similar to those of the horizontal circle. The incremental circle must be referenced before being used. A warning buzzer indicates when this has not been done. Referencing is achieved by rotating the telescope slowly through the zenith. When a certain coding mark on the circle passes a fixed index (detected photoelectronically) the counting begins. The referencing index is set in the factory during manufacture. Deviations of the vertical axis from the vertical are recorded in two directions, one parallel to the telescope axis and the other perpendicular to it and parallel to the trunnion axis, by a tilt sensor. Digital data from the tilt sensor are used to correct automatically the circle readings.

The manufacturers quote a standard error of a horizontal direction and of a vertical angle, of ±0.1 mgon, i.e. $\pm1^{cc}$, or $\pm0.3''$. The maximum rate of rotation of the alidade about the vertical axis is 1.5 revolutions per second.

3.4.4.5 The Aga Industrial Measuring System IMS 1600

This is essentially an Aga Geodimeter 710 electronic tacheometer (the measuring head) which is interfaced with a minicomputer. A printer is also interfaced with the minicomputer. This combination has been designed and programmed to give a rapid graphical and numerical representation of profiles of the ceramic linings of furnaces used for making steel, without the need to shut down the manufacturing process. The combinations of units which comprise the IMS 1600 can be used for similar rapid measurement and analysis in workshops and laboratories, but the IMS 1600 was developed specifically for the iron and steel industry by Aga in conjunction with Domnarvets Järnverk of Borlänge, Sweden.

Fig. 3.42

The IMS 1600 is illustrated in Fig 3.42 which shows the basic Aga Geodimeter 710 (1) with heat shield (2) mounted above the console (3) with protective casing (4). The display/control unit under the flap (5) is used by the operator to instruct the controlling microprocessor of the measuring head and to display horizontal and vertical components of the measured distances and values of horizontal and vertical angles. The minicomputer (6) is a Hewlett-Packard HP 9825A interfaced with the measuring head. This minicomputer transforms the data from the measuring head according to programs stored on magnetic tape cassettes. Programs supplied by the manufacturers for the IMS 1600 are, for the most part, related to the specific requirements of the iron and steel industry, but the user can, of course, develop programs for other purposes. A Hewlett-Packard 9871A printer (7) is interfaced with the minicomputer and is used to print derived data such as internal radii of the furnace and also to depict the data in the form of profiles and sections. In the base (8) of the unit are housed the air filter, fan, transformer and power unit which are necessary for the successful operation of the equipment only a few metres away from an active furnace.

The Aga Geodimeter 710 derives from the Geodimeter 700, aspects of which are described by Rawlinson, 1976. The operating principles of the Geodimeter 710 are similar to those of the earlier Geodimeter 700, but there are important differences. The later model uses a Hughes helium-neon (HeNe) laser to produce a carrier signal of wavelength 632.8 nm. This signal is amplitude modulated at two frequencies (299.7 kHz and 29970.0 kHz) which provide the coarse and fine measurements respectively of the distance. A KDP modulator is used instead of a Kerr cell because turbulence set up in the latter creates 'noise' in the modulation signal. The maximum transmitted power is 300 μW, but the collimation of the beam and the high gain of the avalanche photodiode permit the use of the natural surface of the target as a reflector for EDM.

Figure 3.43 shows a section through the telescope of the measuring head. The modulated signal from the HeNe laser enters the telescope through the trunnion axis and is transmitted to the target through the beam expander lenses (8) and (9). The negative lens (8) can be shifted along the direction of its axis by means of the beam expander control knob (11). This adjustment can be used to concentrate the signal within a circular area only a few millimetres in diameter at targets up to about 20–30 metres distant. This concentration of the signal over a small area is one of the reasons why no artificial reflector is needed as a target. The converging lens (9) of the beam expander combination is fixed to an optical plane plate (1) which also carries a corner-cube prism (4).

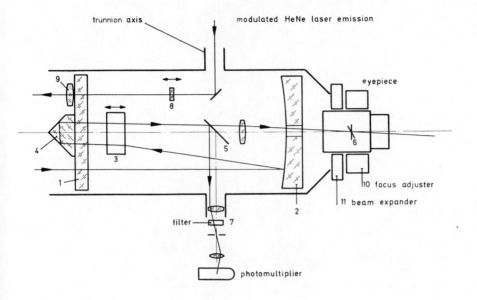

Fig. 3.43

After reflection from the target, the signal enters the telescope through the plate (1) and is reflected from the spherical mirror (2) to the focusing lens system illustrated schematically by (3). The adjustment of this focusing system is made by means of the control knob (10) at the eyepiece. After passing through the lens system (3) the signal is refracted and reflected by the plate (1) and prism (4) to fall on the beam-splitter plate (5). Those components of the signal which have wavelengths in the region of the HeNe laser are reflected by (5) onto an objective lens and a bandpass filter (7) which transmits only the HeNe laser signal to the photomultiplier where it generates a current used for phase comparison and distance measurement. Other components of the incoming signal pass through the beam-splitter plate and produce an image at the reticule (6) which is viewed through the eyepiece of the telescope.

3.4.4.6 The controlling microprocessor of the Keuffel & Esser Vectron

The Keuffel & Esser Vectron surveying system is illustrated in Fig 3.44 and consists of three modules. These are the Vectron instrument which is an electronic theodolite, the Autoranger instrument for EDM and the Vectron field computer for storage, computation and display of data. Each module can be used on its own or in combination with one

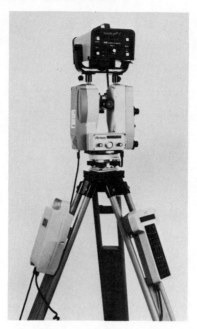

Fig. 3.44

or both of the other two modules. In this section, it is assumed that all three modules are connected.

The controlling microprocessor for data transfer, transformation and storage is illustrated schematically in Fig 3.45. The four units which generate data are shown in the upper left-hand corner and are: the tilt sensor which detects and measures deviations of the vertical axis from the vertical; the horizontal and vertical circle sensors; and the infra-red EDM component. Each of these four devices generates real-time data which are held at the gates G2, G3, G4 and G5 until released by the function enable selector.

The central processing unit (CPU) is an 8-bit microprocessor which uses a 2-byte memory address system. Communications with the locations in the read-only memory (ROM) and random access memory (RAM) are via the address latch (with the higher-order byte) and the address bus. The chip select decoder is used to select the particular RAM or ROM element for communication. It also acts as a binary-coded decimal (BCD) to decimal converter for data to be displayed at the integral display device shown at the top left-hand corner of Fig 3.45. The RAM consists of two (256 × 4) components and the ROM of three (2K × 8) components. The ROM contains the operating program and the RAM is used as an extra scratch-pad for CPU operations during execution of the programs. When the power supply is switched off, data in the RAM are lost.

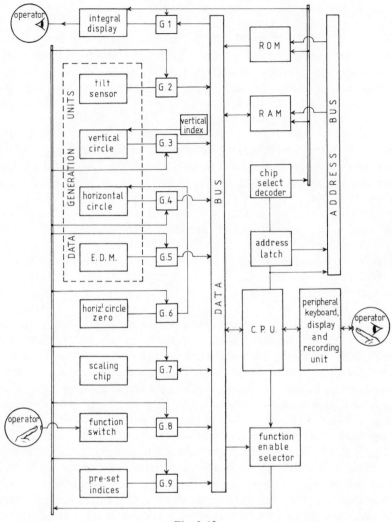

Fig. 3.45

Other components shown in Fig 3.45 are: the vertical circle index and reference devices which set the vertical circle reading to zero whenever the telescope is normal to the direction of gravity as detected by the tilt sensor; the horizontal circle zero which sets the horizontal circle reading to zero when the operator presses the appropriate key; the function switch, which is set by the operator to indicate which data are to be sent to the display; the scaling chip which is a small arithmetic unit capable of scaling the circle readings to the units (centesimal, sexagesimal or mils) selected by the operator's manipulation of a function switch; and the pre-set indices which are defined by laboratory calibrations at the time of manufacture.

To illustrate the way in which the microprocessor controls the registration, transformation and display of data, suppose at a particular time during measurement that the operator wishes to display and record the horizontal circle reading. The program repeatedly interrogates the function switch. When the operator selects 'H' for horizontal angle and say, sexagesimal units, the CPU directs the function enable selector to enable gate G4. Binary data representing the horizontal circle reading pass through the enabled gate G4 and along the data bus to the CPU. The CPU then directs the function enable selector to enable gate G7 so that the binary data representing the horizontal circle reading can pass from the CPU via the data bus to the scaling chip where they are transformed to represent sexagesimal units. These data then pass through enabled gate G7, along the data bus to the CPU which then passes them via the address latch and chip select decoder to the RAM where they are stored. The chip select decoder also transforms the BCD data to decimal data and enables gate G1. These decimal data pass to the integral display where the reading is seen by the operator in sexagesimal units. To record this particular value, the operator signalises this requirement at the peripheral keyboard. In response to this signal, the CPU, via the address latch and chip select decoder, collects the data from the RAM through the data bus and sends them to the data storage locations in the peripheral storage unit. If the function selector switch remains on 'H' and the alidade is rotated about the vertical axis, the microprocessor will continue to present the sexagesimal horizontal circle reading at the integral display, updated at about 0.5 s intervals. Current values will be sent to the peripheral recording unit only when the operator signalises this requirement at the keyboard.

The function switch can be used to obtain and display horizontal and vertical components of measured slope distances, horizontal and vertical circle readings (corrected for residual tilts of the vertical axis) and rectangular components of the measured distance in the horizontal plane ($\triangle E$ and $\triangle N$) if the horizontal circle readings are regarded as bearings.

In the configuration of the Keuffel & Esser Vectron surveying system illustrated in Fig 3.44, the Accuranger EDM component incorporates a microprocessor which transforms digital phase differences between the incoming and outgoing amplitude modulated infra-red signals to distances, corrected for atmospheric refraction and reduced to the projection. In the schematic representation of the controlling microprocessor shown in Fig 3.45, it is assumed that the corrected distance is available from the EDM unit at gate G5. It is part of the purpose of the modular design of the Vectron that each of the three major components (EDM, theodolite and field computer) can be used singly or in any combination, but in an integrated system the CPU would control all the computing steps and transfers of data, including those necessary for EDM.

The field computer has a 40-key alphanumeric keyboard so that names and numbers can be used to identify stations and target positions. It has a solid-state RAM which in its basic form can take 800 'data items', where an item is an angle or a measurement of distance or a five-digit alphanumeric identification. This storage capacity can be extended to 4 800 items. The field computer can also be used for computation of intersections, resections and setting-out data, for example.

3.5 Plate levels

The importance of allowing for non-verticality of the vertical axis when observing a horizontal angle between stations at widely differing elevations is described in section 4.7.1.

A level (the Electrolevel) developed by the British Aircraft Corporation has a much greater sensitivity than the normal plate level and has been used on theodolites for carrying out precise traverses along steep slopes underground.

Figure 3.46 illustrates the principle of the level. The vial is a few centimetres long with a small radius of curvature and precision ground

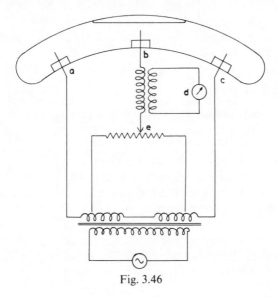

Fig. 3.46

upper surface. It is filled with an electrically conducting fluid similar to the usual alcoholic solution used in plate levels. Three platinum electrodes a, b and c are inserted into the vial. Operation depends upon the fact that the electrical resistance of the fluid between a pair of electrodes varies with the inclination of the vial. The two portions of the fluid ab and bc form two branches of a Wheatstone bridge arrangement subjected to an alternating current of 400 Hz. The input voltage is 1 volt and the overall resistance about 9000 ohms.

As the inclination of the vial is changed, the bridge becomes unbalanced with a resultant indication on the meter d. This can be nulled by means of the variable resistance e, which forms the other two branches of the Wheatstone bridge.

The deflection of the meter d can be calibrated to represent the angle of tilt. A sensitivity of 1 second of arc per scale division is possible with estimation to 0.25″. Such a high sensitivity would make the coarse levelling-up procedure impossible. Accordingly, by suitable amplification, the sensitivity of the meter can be switched to 10 seconds of arc per division over a range of a few minutes. The deflection of the needle is instantaneous and damping is effective within one second of time.

This level has many applications in industry where it is used to detect movements of machine tools and deflections in structures.

The Talyvel electronic level manufactured by the Rank Organisation is rather more bulky than the Electrolevel and therefore not as suitable for attachment to a theodolite. A pendulum carrying an armature is suspended between the two halves of a transducer which in turn are attached to the plane whose inclination is being measured. A tilt causes the pendulum to move nearer to one half of the transducer thereby producing a displacement signal. The pendulum is liquid-damped. The sensitivity can be as much as 1 second of arc per division, i.e. the same as that of the Electrolevel.

3.6 Centring systems

In section 2.6.5, centring systems are described in general terms and two types identified: self-locating and stub axis with lateral stop.

As an example of the first type, Fig 3.47a shows the Wild three stud method, and Fig 3.47b shows the Kern system. As examples of the stub axis with lateral clamp systems, Fig 3.48a shows a Zeiss (Oberkochen) 34 mm stub axis and Fig 3.48b illustrates the short conical stub axis used by Watts. Results of accuracy tests are given in section 5.6.

Fig 3.49 shows the Kern levelling and centring system whereby the tripod head (4) need be set up only approximately in a horizontal plane. A plumbing rod (2) placed vertically over the station mark

Fig. 3.47a

Fig. 3.47b

according to the bubble (1) connects with a plate (5) resting on a dome (3) which can slide across the tripod head. When the foot of the plumbing rod is on the station mark and the rod is vertical, the plate (5) and dome (3) can be clamped to the tripod head by the clamp (6). In this situation, the plate (5) is nearly horizontal and the instrument can be located on the plate and will be centered and almost levelled. The 'footscrews' on the Kern instruments have horizontal axes; the knobs connect with cams which tilt the vertical axis into the required position. The 'footscrews' need have only short runs since plate (5) is always nearly horizontal for each set-up.

Another interesting feature designed to speed-up the centring and levelling process is the parallel-motion device of Zeiss (Oberkochen) illustrated in Fig 3.50.

Two studs in the tripod head are located in a pair of parallel slots in the base-plate. The three studs in the tribrach or levelling head are located in three slots at right angles to the previous pair of slots. Thus movement of the theodolite relative to the tripod head can take place

Fig. 3.48a

Fig. 3.48b

in either of these mutually perpendicular directions and the motion will be irrotational.

By constraining the theodolite to move without rotation, initial levelling will not be disturbed by subsequent centring as is usual with below-the-footscrews centring. Thus the method gives the compactness of the system without the extra time necessary when levelling with the optical plummet.

3.7 Automatic vertical circle indices and tilt sensors

The difference between automatic indexing of the vertical circle and referencing of an incrementally coded vertical circle has been described in sections 2.7 and 2.8. Referencing is carried out to ensure

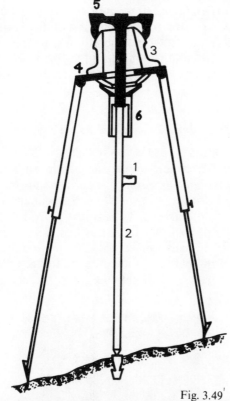

Fig. 3.49

Fig. 3.50

that the circle is read as zero (or 90° exactly) when the telescope is perpendicular to the vertical axis of the theodolite. Indexing is necessary to ensure that the circle is read as zero (or 90° exactly) when the telescope is horizontal. Referencing is internal, but indexing is external because it is related to the direction of gravitational force at the instrument. In some electronic theodolites, referencing is carried out by the operator after setting-up and by relying on the automatic indexing. In others, referencing is carried out in the factory during manufacture.

3.7.1 The automatic index of the Kern K1A Engineer's Theodolite

The automatic pendulum index on the Kern K1-A theodolite is illustrated in Fig 3.51, which shows the pendulum support (1) the pendulum arm (2) the air damping cylinder (3) and the objective (4).

Figures 3.52(a and b) illustrate the system. Suspended from pivot (1) is a lens (4) attached to a frame (2). The distance from the pivot to the lens is equal to the radius of the circle. In Fig 3.52a, the vertical axis is vertical and the telescope is horizontal. On the assumption that the reading system is in the plane of the circle, an image of the zero mark (V) on the circle is formed at the reticule (6) by the suspended lens (4) and the fixed lens (5).

Fig. 3.51

Suppose now that the vertical axis is inclined by an angle e in the plane of the circle (Fig 3.52b), but that the telescope remains horizontal.

If no automatic index were present, lenses (4) and (5) would form an image at the reticule (6) of the circle reading at X. However, because of the fact that lens (4) is suspended vertically below the point of suspension (1) and at a distance from it equal to the radius of the circle, the collimating ray is deviated through an angle e. Thus an image of the zero of the circle is formed at the reticule, this being the correct reading.

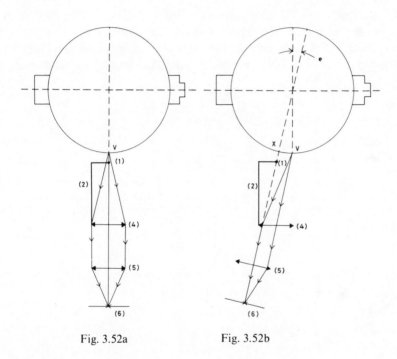

Fig. 3.52a   Fig. 3.52b

3.7.2 The tilt sensor of the Hewlett-Packard 3820A Electronic Total Station

The principle underlying the use of the tilt sensor is described in section 2.7.3 and its application in the Hewlett-Packard 3820A is described by Gort, 1980.

Figure 3.53 shows a vertical section through one of the standards. Radiation from an electroluminescent diode passes through a lens system and falls with uniform intensity on a sinusoidal slit pattern similar to that used on the main circles of the instrument. An image of the slit pattern is produced at the array of photodiode sensors by refraction through two lens systems and reflection from a mercury pool damped by silicone oil. The effective focal length of the imaging system is 163 mm.

If the vertical axis is displaced by an angle e radians in the vertical plane through the slit pattern, the image of the pattern will be displaced by 326e at the photodiodes. Four photodiodes equally spaced over three-quarters of a period of the sinusoidal pattern measure this displacement in the way described in section 3.4.4.3.

Two sets of slit patterns and diode arrays arranged orthogonally are used as shown in Fig 3.54 so that tilts of the vertical axis in the direction of the line of sight and in the direction of the trunnion axis can be measured. The controlling microprocessor automatically computes and applies corrections to the horizontal and vertical circle readings to offset the effects of the tilt of the vertical axis. It is therefore sufficient to set up the instrument so that the vertical axis is only

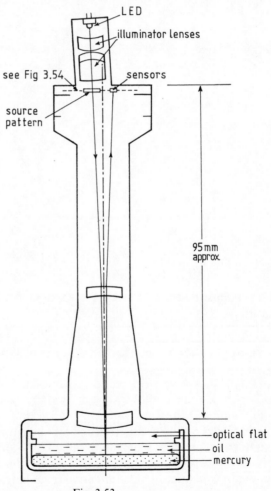

Fig. 3.53

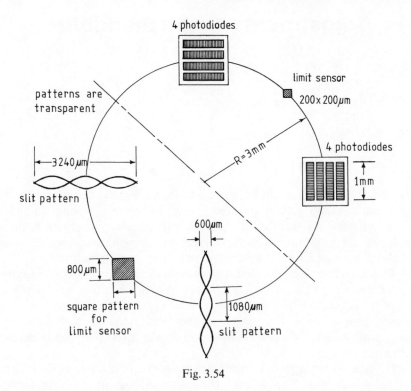

Fig. 3.54

approximately vertical. A conventional tubular plate level is not necessary and a spherical spirit level is sufficient to ensure that the vertical axis is set within the range of the tilt sensor. This range is about 2′ (or 4ᶜ). If the vertical axis is within the range, illumination from the square pattern (Fig 3.54) will fall on the limit sensor. If the vertical axis is outside the range of the tilt sensor, no radiation falls on the limit sensor and a visible warning is given to the operator. The maximum error arising from the measurement of the inclination of the vertical axis to the vertical is about 1″ or 3ᶜᶜ (Gort, 1980).

4 Adjustments to the theodolite

It is not possible to make a theodolite in which the relationships between axes and other components are correct and remain so under all conditions. Therefore it is necessary to consider what effects these mechanical and optical defects have on measurements made with theodolites and to what extent these effects can be reduced.

The defects can be divided into two classes; defects of construction and defects of adjustment. Defects of construction (malconstructions) cannot, in general, be put right by the surveyor in the field, although in many cases the observations can be arranged so that these defects have negligible effects. Malconstructions and their effects are considered in chapter 5.

Defects of adjustment (maladjustments) can be put right by the surveyor in the field, and the instructions in manufacturers' handbooks should always be followed. Although theodolites differ, there are basic principles of construction and adjustment which apply to most models and the following descriptions are in general terms.

4.1 Vertical axis and plate level adjustments

These adjustments are made to ensure that the vertical axis of the theodolite is vertical. This is most easily achieved if the principal

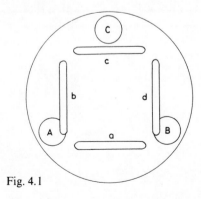

Fig. 4.1

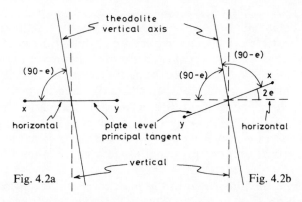

Fig. 4.2a   Fig. 4.2b

tangent of the plate level is perpendicular to the vertical axis of the theodolite, although verticality of the vertical axis can be achieved with an oblique principal tangent provided the obliquity does not exceed about 1′.

The following description is based on the assumption that there is a small error (e) in the inclination of the principal tangent relative to the vertical axis.

1. With the theodolite attached to a firm tripod, clamp the lower plate (if the theodolite has a lower plate clamp), unclamp the upper plate and rotate the alidade until the plate level, a, is parallel to the line joining any two footscrews (A and B for example in Fig 4.1).
   Bring the bubble to the centre of the vial by turning footscrews A and B by equal amounts in opposite directions (the bubble 'follows' the left thumb).
2. Rotate the alidade through 90° until the plate level is in position b (Fig 4.1) and centre the bubble using the third footscrew, C, only.
3. Repeat steps 1 and 2 until the bubble is central in both positions.
4. Rotate the alidade until the plate level is 180° from its original position (position c, Fig 4.1). If the bubble remains central, the adjustments are complete; the vertical axis is vertical and the principal tangent of the plate level is perpendicular to it. If the bubble moves off centre, the adjustments are not complete; the vertical axis is not vertical and the principal tangent of the plate level is not perpendicular to it. Figures 4.2a and 4.2b illustrate the relationships between the axes and horizontal and vertical directions in vertical planes after steps 1 and 4 respectively (assuming an error e in the inclination of the principal tangent of the plate level to the vertical axis of the theodolite).
5. When the plate level is in position c (Fig 4.1) it can be seen from

Fig 4.2b that the vertical axis is at an angle e to the vertical whereas the principal tangent of the plate level is at an angle 2e to the horizontal. The main purpose of this adjustment is to ensure verticality of the vertical axis and this is done by bringing the bubble back half-way towards the centre of the vial (measured against the graduations on the vial) using footscrews A and B (Fig 4.1). The relationships between the axes and horizontal and vertical directions in a vertical plane parallel to that through the centres of A and B are then as shown in Fig 4.3a; the vertical axis is vertical, but the principal tangent of the plate level is still in error by an angle e.

6. Rotate the alidade until the plate level is 270° from its original position (position d in Fig 4.1). Using footscrew C bring the bubble to the same position in the vial as that reached after step 5. This ensures that the vertical axis is vertical in a plane perpendicular to that of Fig 4.3a. The bubble should now remain in the displaced position for all pointings of the alidade; the vertical axis is vertical, but the principal tangent of the plate level is not perpendicular to it. This maladjustment can be corrected if desired after step 5 above. The procedure will then be as follows:

5a Bring the bubble to the centre of the vial by means of the capstan screws or nuts, giving the situation in Fig 4.3b.

5b Repeat steps 1–4. The bubble should now remain central for all pointings of the telescope; the vertical axis is vertical and the principal tangent of the plate level is perpendicular to it.

From the foregoing it is apparent that orientation of the vertical axis can be carried out with an oblique plate level provided that the displacement of the bubble after placing it in position c (Fig 4.1) is still within the vial graduations; the displacement can then be measured and half the error taken out. If the displacement of the bubble is large and the bubble is outside the vial graduations then the displacement

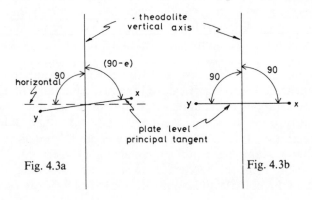

Fig. 4.3a   Fig. 4.3b

must be estimated and the procedure repeated until the displacement is small enough to be measured against the graduations.

4.2 Adjustment of the line of collimation
(Horizontal collimation adjustment)

This adjustment should follow the test and adjustment of the vertical axis and plate level described in section 4.1.

The requirement is for the line of collimation to be perpendicular to the trunnion axis. There are two different methods of carrying out this adjustment; one is independent of the horizontal circle and therefore independent of circle graduation and reading errors and the other is not, but it does depend upon the accurate placing of chain arrows and is likely to be less accurate than the second method in the case of a modern theodolite. The first method depends upon the fact that rotation of the line of collimation about the trunnion axis results in the tracing-out of a cone if the line of collimation is not perpendicular to the trunnion axis. The procedure is as follows.

1. Select an open, level site about 200 m long and set-up and level the theodolite in the centre of the open stretch, at T in Fig 4.4a. The trunnion axis is t_1t_2 and the collimation error is i_c.
2. Bisect a suitable target (a chain arrow for example) with the vertical hair and clamp all horizontal movements. Let this target be A in Fig 4.4a. Eliminate parallax (section 4.5) and bisect A accurately.

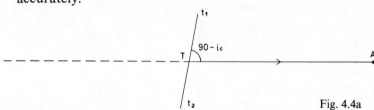

Fig. 4.4a

3. Transit the telescope and mark point B with a chain arrow, on the line of collimation, such that TB $\simeq$ TA (Fig 4.4b).

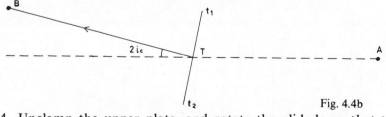

Fig. 4.4b

4. Unclamp the upper plate, and rotate the alidade so that the vertical hair again bisects target A (Fig 4.5a). Clamp all horizontal movements.

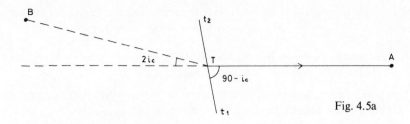

Fig. 4.5a

5. Transit the telescope and mark point C with a chain arrow on the line of collimation so that TC ≏ TB (Fig 4.5b).

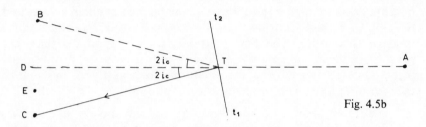

Fig. 4.5b

6. Measure BC and place arrows at D (BD = CD) and at E (CE = 1/4 BC).
7. Move the reticule horizontally by means of the locking screws until the vertical hair passes through E. This must be done with care, the screw towards which the reticule is to move must be slackened first. Finally, tighten the screws carefully.
8. Repeat steps 1 to 3 and check that the vertical hair now bisects the arrow at D. If not, repeat the test and adjustment until it is satisfactory.

The procedure in the second method is as follows:

1. Set up the theodolite and level it as described in section 4.1. Clamp the lower plate.
2. Select a distant target at roughly the same level as the theodolite and which can be easily bisected. Eliminate parallax (section 4.5). Bisect the target with the vertical hair on F.L. and read the horizontal circle.
3. Repeat step 2 on F.R.
4. The two readings should differ by exactly 180°. But suppose the readings are: from step 2 (F.L.) 209° 42′ 12″ and from step 3 (F.R.) 29° 41′ 40″. The difference is 180° 00′ 32″. The inclination of the line of collimation to the trunnion axis is therefore in error by 1/2 (32″) i.e. by 16″ (this is the horizontal collimation error) and the readings should be:

F.L. 209° 42′ 12″ − 16″ = 209° 41′ 56″)

and:

F.R. 29° 41′ 40″ + 16″ = 29° 41′ 56″).

5. With the theodolite still in the F.R. attitude, set 29° 41′ 56″ on the horizontal circle using the micrometer and upper plate tangent screw. This will throw the vertical hair off the target.
6. Move the reticule horizontally by means of the locking screws until the vertical hair again bisects the target.
7. Repeat steps 2 and 3 on a different horizontal circle setting. If the F.L. and F.R. readings differ by 180° 00′ 00″ ± 2e, where e is less than 5″ for a 1″ theodolite (or 30″ for a 20″ theodolite) the adjustment can be considered satisfactory.

4.3 Orientation of the reticule

The previous adjustment involves moving the reticule. Therefore it is necessary subsequently to check whether the vertical hair is in a plane perpendicular to the trunnion axis. The procedure is as follows.

1. Set up and level very carefully.
2. Select a target similar to that for the previous adjustment and bisect it with the vertical hair, on either face.
3. Using the vertical tangent screw, rotate the line of collimation about the trunnion axis. If the vertical hair stays on the target, then the adjustment is satisfactory.
4. If the hair moves off the target, loosen the reticule locking screws and rotate the reticule until the vertical hair stays on the target as the telescope is rotated about the transit axis. Tighten the locking screws.
5. Repeat the test until the adjustment is complete.

If this adjustment is carried out, the previous adjustment (section 4.2) must be re-checked and any necessary correction made. On completion of the two adjustments the line of collimation will be perpendicular to the trunnion axis and the vertical hair will be in a plane perpendicular to the trunnion axis. The manufacturer will have ensured that the horizontal hair is perpendicular to the vertical hair.

4.4 Adjustment of the vertical circle index
(Vertical collimation adjustment)

This must always be carried out if the reticule has been adjusted as described in sections 4.2 and 4.3. The purpose of the adjustment is to ensure that the vertical circle reading indicates zero when the line of collimation is horizontal. In most modern theodolites the vertical circle rotates with the telescope whilst the circle reading index remains stationary. There is usually no provision for moving the circle in relation to the telescope so the adjustment has to be carried out on the index.

As with the adjustment described in section 4.2, there are two methods, one independent of the circle graduations (and their errors) and the other not. A description of the latter method is described here, because circle graduation errors of a modern theodolite are unlikely to affect the adjustment adversely.

1. Set up and level very carefully.
2. Select a distant target a few degrees above or below the horizontal which can be accurately bisected. Eliminate parallax (section 4.5). Bisect the target with the horizontal hair on F.L. Centre the bubble in the altitude level very carefully and read the vertical circle.
3. Repeat step 2 on F.R.
4. The vertical angle on F.R. should be equal to the vertical angle on F.L. if there is no index error. If the angles are not equal, the mean value is correct and the index error is the difference between this value and either of the two measured values. Suppose the following readings were obtained:

$$\text{F.L. } 92° \ 14' \ 30''$$
$$\text{F.R. } 267° \ 43' \ 48''$$

These readings correspond to the vertical angles

$$\text{F.L. } -2° \ 14' \ 30''$$
$$\text{F.R. } -2° \ 16' \ 12''$$

The mean value is $-2° \ 15' \ 21''$ and the index error is $51''$.

5. With the theodolite still in the F.R. attitude, calculate the circle reading corresponding to the correct angle and set this on the circle using the micrometer and the altitude level setting screw. In the example, the F.R. circle reading to be set is $270° \ -2° \ 15' \ 21'' = 267° \ 44' \ 39''$.
6. This will cause the altitude level to run off centre. Bring it back to its central position by means of the capstan headed adjusting screws.
7. Repeat the test using a different target until the collimation error is within the limits stated in section 4.2.

The foregoing description is applicable to a theodolite having an optical micrometer for reading the vertical circle and an altitude level and setting screw for indexing the circle. The test for a theodolite with an automatic index follows steps 1–4 above, inclusive (without the need to set the index manually in step 2). Once the vertical collimation or index error has been determined, the adjustment should be made according to the manufacturer's instructions. This usually means setting the micrometer to give the correct reading and then adjusting the index by turning an adjusting screw until the coincidence of the main-scale graduations is achieved. Again, the adjustment should be followed by an independent test using a different target.

4.5 Focusing and elimination of parallax

These are adjustments to the telescope which are just as necessary as the other adjustments if the theodolite is to give the results of which it is capable. Their purpose is firstly to adjust the eyepiece so that the image of the reticule is between the observer's near point and far point (see section 2.1.1.) and secondly to place the intermediate image (section 2.1.7, Fig 2.4) in the plane of the reticule.

4.5.1 Adjustment of the eyepiece

If the eyepiece is gradually screwed out from its innermost position, an observer will notice the following sequence of events. Initially the cross-hairs cannot be seen. Then they appear and gradually become blacker and better defined. They are sharply defined for a short time and then become blurred and less black, finally disappearing altogether when the eyepiece is at its maximum distance from the reticule.

As the eyepiece is withdrawn, the image of the cross-hairs moves away from the observer's eye. When the image is at the near point, it is sharp and stays sharp as it moves towards the far point. It then becomes blurred as it passes the far point. For the most comfortable viewing conditions, the image of the cross-hairs should be at the observer's far point as the ciliary muscles will then be relaxed.

The procedure for achieving this is therefore as follows:

1. Point the telescope to the sky or obtain even illumination of the field of view by holding a hand a short distance in front of the objective.
2. Starting with the eyepiece in its innermost position, gradually withdraw it until the cross-hairs are seen clearly.
3. Continue to withdraw the eyepiece until the sharp image just begins to become blurred.
4. Screw the eyepiece in very slightly to obtain the sharp image again. Note the reading on the eyepiece focusing ring. This can then be set each time the theodolite is used.

This adjustment (and all observations with theodolites and levels) should be made with both eyes open, this being less tiring than keeping one eye shut, although it requires practice.

4.5.2 Adjustment of the intermediate image
(Elimination of parallax)

If the intermediate image of the target formed by the objective and the focusing lens is placed in the plane of the reticule, then it will be seen easily and distinctly by the observer as a result of the previous adjustment.

The procedure for each pointing is as follows:

1. Obtain the target in the field of view.
2. Keeping the eye fixed on the cross-hairs, adjust the focussing screw until the image of the target becomes sharp.
3. Carry out the final adjustment of the focusing ring so that there is no relative motion between the image of the target and the image of the cross-hairs as the eye is moved slightly from side to side.

When there is no relative motion between the target and the cross-hairs, they must be in the same plane. It is not possible to ensure coincidence between the image of the target and the cross-hairs simply by adjusting the focusing lens to give a sharp image of the target; the eye can accommodate two images separated by a small distance. Therefore the distance between the image of the target and the cross-hairs (i.e. parallax) must be eliminated by testing for relative motion. Bisection of the target by the cross-hairs will then be independent of the position of the observer's eye.

This adjustment should be carried out for each pointing, although if the targets are all distant, elimination of parallax for one of them will also eliminate parallax for the others; it is only when the targets lie at short (i.e. under 30 m) distances from the theodolite that adjustment for each pointing will be found necessary.

4.6 Adjustment of the optical plummet

The method will depend upon whether the plummet optics are situated in the alidade (e.g. Vickers Tavistock) or in the tribrach (e.g. Wild T2).

If the former, the field method is to set-up over hard flat ground and level carefully. Fix a piece of paper on the ground below the instrument and make a mark on it to coincide with the optical axis of the plummet. Rotate the alidade through 180° about the vertical axis. The plummet axis will also be rotated and if itself is vertical, will still pass through the mark on the paper. If it is out of adjustment however, it will give a new intersection with the paper which should be marked. A point mid-way between the two marks is the point through which the plummet axis should pass, and this should also be marked.

In order to deviate the plummet axis so that it passes through the correct mid-position, adjusting screws on either the objective or the reticule of the plummet should be used according to the manufacturer's handbook. The test should be repeated until the adjustment is satisfactory.

If the plummet is in the tribrach, the previous method cannot be used because the optical axis of the plummet cannot be rotated through 180° without disturbing the levelling.

One way of carrying out the adjustment in the field is to suspend a plumb bob from the theodolite and then to set the plummet against the plumbed mark. However this method is not really suitable since the plummet should be more accurate than the plumb bob. An alternative method is to set-up the theodolite above a dish of mercury, thereby making use of the principle of auto-collimation. The image of the centre of the plummet objective should coincide with the intersection of the cross-hairs on the plummet reticule. If it does not, then the plummet is adjusted until coincidence is obtained. Careful levelling is important.

Another method which does not require a dish of mercury is to clamp the telescope roughly horizontal and then to place the theodolite on a table so that it rests on the vertical circle casing, the other standard and the telescope objective. In this position the theodolite should be supported so that its vertical axis is more or less horizontal and so that the tribrach can be rotated about the vertical axis relative to the alidade. The test and adjustment can then be carried out as described earlier, but this time the intersections of the optical axis with a card attached to the wall are marked and the plummet adjusted to the central position. The card should be approximately normal to the axis of the plummet.

4.7 Effects of maladjustments on observations

It is apparent from the introductory paragraphs to this chapter and from the sections which followed that it is impossible for a theodolite to be in perfect adjustment. Even if the adjustments have been carried out, residual errors will still be present. Their effects on observations are now described.

It is convenient to consider the theodolite as being at the centre of a very large sphere, large enough for the theodolite dimensions to be negligible, and to examine the intersections of the axes of the theodolite with the surface of the sphere. Thus Fig 4.6 illustrates these intersections for a theodolite in perfect adjustment.

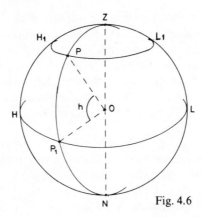

Fig. 4.6

The theodolite is at O, the centre of the sphere. The vertical axis is vertical and intersects the sphere at the zenith– and nadir–points Z and N respectively.

When the line of collimation is directed towards a given point P at an angle of elevation h, rotation about the vertical axis results in the trunnion axis tracing out the horizontal great circle $HP_1 L$ and the line of collimation tracing out a cone, intersecting the sphere in the horizontal small circle $H_1 PL_1$. Rotation of the line of collimation through P about the trunnion axis results in the vertical great circle $ZPP_1 N$.

To derive relationships between the elements of a spherical triangle, the rules of elementary spherical trigonometry are used. These rules and their application are described in elementary textbooks on the subject (for example Todhunter & Leatham, 1956) and the rules are not given here. In particular, for a right-angled spherical triangle, Napier's analogies are a convenient means of deriving the appropriate relationship for a particular configuration of theodolite axes and spatial directions.

4.7.1 Effect of non-verticality of the vertical axis

Let the inclination be i_v in Fig 4.7a ($ZZ' = i_v$). OZ is the true vertical but OZ' is the direction of the vertical axis. Rotation about this inclined vertical axis results in the trunnion axis tracing out the great circle $H'P_1'L'$ inclined at an angle i_v to the horizontal plane $HP_1 L$. The line of collimation traces out the great circle $Z'PP_1'P_1''$ when rotated about the trunnion axis.

The error in azimuth, α, of point P is $(LP_1'' - LP_1) = P_1 P_1''$ which is e_v say. In order to obtain an expression for this error in terms of i_v, α and h (the vertical angle to P) consider first the spherical triangle $ZZ'P$ (Fig 4.7b). In any spherical triangle ABC,

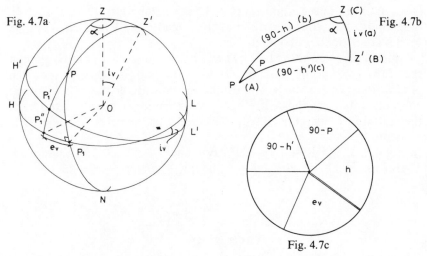

Fig. 4.7a

Fig. 4.7b

Fig. 4.7c

$$\cot a \sin b = \cot A \sin C + \cos b \cos C.$$

Thus for triangle ZZ'P:

$$\cot i_v \sin (90 - h) = \cot P \sin \alpha + \cos (90 - h) \cos \alpha$$

therefore

$$\cot P = \frac{\cot i_v \cos h - \sin h \cos \alpha}{\sin \alpha}$$

Now consider the right-angled spherical triangle PP_1P_1'' (Fig 4.7c)

$$\sin h = \tan e_v \tan (90 - P)$$

$$\frac{\tan e_v (\cot i_v \cos h - \sin h \cos \alpha)}{\sin \alpha}$$

therefore $\tan h \sin \alpha \tan i_v = \tan e_v (1 - \tan h \cos \alpha \tan i_v)$

so $\qquad i_v'' \tan h \sin \alpha \doteqdot e_v'' (1 - \tan h \cos \alpha \, i_v''/\rho'')$

and $\qquad e_v'' \doteqdot i_v'' \tan h \sin \alpha$, because $e_v''.i_v''$ is second-order.

This relationship has been derived on the assumption that the horizontal circle remains horizontal when the vertical axis is inclined to the vertical. This is not so; the horizontal circle will be inclined at an angle e_v to the horizontal and will lie in the plane H'OL' (Fig 4.7a). The effect of this additional error is shown in section 5.2 to be negligible.

Change of face on P will not result in elimination of this error; the line of collimation will still trace out the great circle $Z'PP_1'P_1''$ when rotated about the trunnion axis. Therefore if the effects of this error are to be reduced a measure of the inclination of the vertical axis has to be obtained from the plate level and the correction calculated. The plate level is adjusted (section 4.1) so that its principal tangent is perpendicular to the vertical axis. It therefore gives a measure of $i_v'' \sin \alpha$

(= I_v'') for use in the equation $c'' = I_v'' \tan h$, where c'' is the correction in seconds of arc to the horizontal circle reading to P and h is the angle of elevation or depression to P.

If, for a given pointing to P, the left-hand end of the bubble is against the Lth graduation and the right-hand end is against the Rth graduation (where the graduations are considered to be numbered consecutively from the centre outwards) then I_v can be calculated from the equation $I_v'' = [(L - R)/2] v''$, where v'' is the value of one bubble division.

In Fig 4.8, L = 1 R = 4 v = 20'' say, hence $I_v'' = -3/2 \times 20 = -30''$
Therefore, if h = $-20°$, c = $-30''$ tan $-20°$ = $+10''.8$

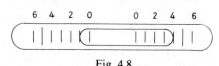

Fig. 4.8

If several pointings are made to P and a mean taken then this mean can be corrected from the equation

$$c'' = \frac{(\Sigma L - \Sigma R)}{2n} v'' \tan h$$

where n is the number of pointings made and ΣL and ΣR are the sums of all the left-hand and right-hand bubble readings respectively.

If v is not known (generally it is engraved on the vial) it can be found by the method described in section 4.7.1.1. This correction is usually only appreciable when a pointing is made to a target at a considerable angle of elevation or depression, for example in astronomical observations or in carrying a traverse up or down a steep slope.

As far as vertical angles are concerned, non-verticality of the vertical axis will not affect the readings significantly. The reason for this is that the horizontal reference is established for each pointing by setting the vertical circle index according to the altitude level. Provided this level is in adjustment (section 4.4) the horizontal datum is correctly established independent of the vertical axis attitude. The error arises from the inclination of the vertical circle to the vertical, which is shown in section 5.2 to be negligible.

Note that the effect of inclination of the vertical axis on a horizontal circle reading depends on α and that if P happens to lie in the same direction as the tilt of the vertical axis then there is no error in the horizontal circle reading from this source.

4.7.1.1. Determination of the bubble value of the plate-level

It may be necessary to replace a broken plate level with one which does not have the bubble value engraved on it. If it is subsequently found necessary to correct observations for non-verticality of the vertical axis then the bubble value must be known. The procedure for determining this bubble value is as follows:

1. Set-up the theodolite on a firm tripod, make the vertical axis vertical and turn the alidade until the telescope is pointing over one footscrew. Clamp all movements and read the horizontal and vertical circles, remembering to centre the altitude level.
2. By turning that footscrew lying under the telescope through about one revolution tilt the vertical axis. This should not affect the plate level. If it does, the telescope has not been correctly aligned over the footscrew, and it must be adjusted.
3. Read the vertical circle again. The difference between this reading and the previous reading gives i, the inclination of the vertical axis to the vertical in the vertical plane passing through the centre of the instrument and the centre of the footscrew.
4. Using the upper plate slow-motion screw, turn the alidade in azimuth until the plate level bubble has moved through n divisions. Read the horizontal circle. The difference between this reading and the previous reading is α, the angle turned through.

If v is the value of one bubble division then $\tan \delta = \sin \alpha \tan i$ where $\delta = nv$. Hence v can be found. This relation can be derived from Fig 4.9. After adjustment of the relevant footscrew, the vertical axis lies along OZ' at an angle i to the vertical OZ. The principal tangent of the plate level will then lie along OB_1. After rotation of the alidade through α about OZ', the principal tangent of the plate level will lie along OB_2, inclined at an angle δ (= nv) to the horizontal. In fact, the angle α measured is angle B_2OB_1, but it is shown in section 5.2 that the error introduced by putting $\alpha = < COB_1$ is negligible.

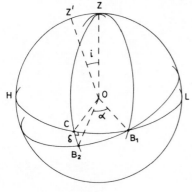

Fig. 4.9a

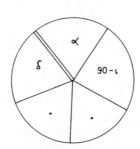

Fig. 4.9b

From the right-angled spherical triangle B_1CB_2, (Fig 4.9b)

$$\sin \alpha = \tan \delta \cot i$$

therefore $\qquad \tan \delta = \sin \alpha \tan i$

A method for determination of the sensitivity (or bubble value) of level bubbles on astronomical and other high-order theodolites is described by Baldini, 1975.

It is worth while reminding the reader that if the plate level is found to be broken in the field the altitude level (if one is fitted) can be used to set the vertical axis vertical following the procedure normally employed using the plate level. The altitude level setting screw should be used to bring the bubble back half-way to its central position instead of using the bubble adjusting screws; this would disturb the vertical circle index setting. A full description of a procedure using the altitude level instead of the plate level to set accurately the vertical axis of a theodolite is given by Bennett & Groenhout, 1976, who also describe a procedure using an automatic index to achieve accurate setting of the vertical axis. Jackson, 1980 also describes a method which uses the automatic index instead of the plate level.

4.7.2 Effect of the line of collimation being inclined to the trunnion axis

Let the line of collimation be inclined at an angle i_c to the normal to the trunnion axis, (see Fig 4.10a). Rotation of the line of collimation about the trunnion axis will result in the tracing out of a flat cone of semi-apex angle $(90 - i_c)$ intersecting the imaginary sphere in the small circle $Z'PP_1'$. ZPP_1 is the vertical great circle through P. ZN is part of the vertical great circle parallel to the small circle $Z'P_1'$. The error in azimuth is $P_1N = e_c$ say.

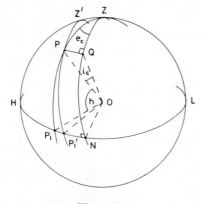

Fig. 4.10a

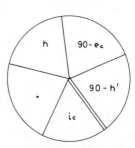

Fig. 4.10b

Let PQ be the arc of a great circle through P perpendicular to ZN.

Then $PQ = i_c$

From the right-angled spherical triangle ZQP (Fig 4.10b)

$$\sin i_c = \cos h \sin e_c$$

therefore $$\sin e_c = \sin i_c \sec h$$

so $$e_c'' \simeq i_c'' \sec h$$

By changing face on P, the small circle is transferred to the opposite side of the great circle ZPP_1, the inclination will be of the same magnitude but opposite in sign, so the mean of the F.L. and F.R. pointings will be free from this error.

If a horizontal angle between two points is measured on one face only, then that angle will be free from the error only if the vertical angle is the same to each point.

The error in a vertical angle is the difference between h and arc NQ $(= h')$. From the right-angled spherical triangle ZQP (Fig 4.10b)

$$\sin h = \sin h' \cos i_c$$

But i_c will be at most a few minutes, so $\cos i_c \simeq 1$ and for all practical purposes $h = h'$.

4.7.3. Effect of incorrect orientation of the reticule

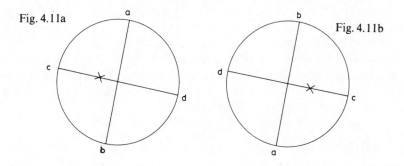

Fig. 4.11a

Fig. 4.11b

On F.L., Fig 4.11a illustrates the bisection of a target X by the 'horizontal' hair cd. If on change of face, the target is again bisected by the same part of the hair (Fig 4.11b) the mean of F.L. and F.R. will be free from this error. Obviously the error on any one face will be nil if bisection is made by the junction of the hairs, but this is often inconvenient in practice.

Similar reasoning applies to bisection by the 'vertical' hair ab.

4.7.4 Effect of a vertical circle index error

This arises when the principal tangent of the altitude level is not parallel to the line joining the vertical circle index to the centre of the graduations of the vertical circle.

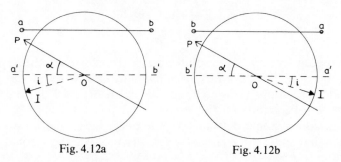

Fig. 4.12a   Fig. 4.12b

In Fig 4.12a, ab is the principal tangent of the altitude level (adjusted to be horizontal by the altitude level setting screw).

O is the centre of the vertical circle graduations,

I is the reading index,

a'b' is a line though O parallel to ab,

i is the inclination of the line a'b' to IO,

and α is the true vertical angle to a point P.

The reading on the circle is $(\alpha + i)$. On changing face (Fig 4.12b) the vertical angle is read as $(\alpha - i)$. Therefore the mean of F.L. and F.R. is α, the true vertical angle.

Although in theory a large error i can be tolerated, in practice it should be made small by the adjustment described in section 4.4 so that the surveyor can see whilst observations are being carried out whether a gross error has occurred in his readings.

4.7.5 Effects of parallax

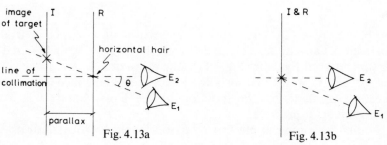

Fig. 4.13a   Fig. 4.13b

In Fig 4.13a, R is the plane of the reticule and I is the plane of the intermediate image which has been incorrectly formed away from the plane of the reticule. With the eye in position E_1 the image of the target appears to be on the horizontal hair, but with the eye at E_2, on the line of collimation, bisection is not obtained. Thus an incorrect pointing can be made from E_1 and the angular error (θ) is inversely proportional to the distance of the target. It can amount to a few minutes of arc for a badly adjusted focusing lens. The elimination of parallax (Fig 4.13b) is therefore essential for accurate bisection. Although it is a random error, generally insufficient observations to a target are taken for its effects to be considerably reduced in the mean. Failure to eliminate parallax is one of the major causes of errors by inexperienced observers. The procedures outlined in section 4.5.1 and 4.5.2 should always be followed.

4.8 Effect of miscentring

Suppose a horizontal angle ABC (Fig 4.14) is to be measured, but owing to miscentring the theodolite is set-up over B′ instead of B. Then the error in the horizontal reading to A will be (putting BB′ = e and AB = D):

$$\alpha'' \doteqdot e \sin \theta \; \rho''/D$$

This will be a numerical maximum when $\theta = 90°$ or $270°$ when $\alpha'' \doteqdot e\rho''/D$.

The increased importance of accurate centring over short distances is obvious.

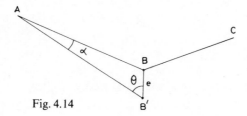

Fig. 4.14

4.9 Summary

Type of maladjustment	Adjustment	Effect of maladjustment	Action taken
1. Vertical axis not vertical	Section 4.1	error in horizontal circle reading = $e_v = i_v \tan h \sin \alpha$	Calculate correction $\dfrac{\Sigma L - \Sigma R}{2n}\ v \tan h$
		error in vertical circle in reading negligible	
2. Line of collimation not perpendicular to trunnion axis	Section 4.2	error in horizontal circle reading = $e_c = i_c \sec h$	Take mean of F.L. and F.R.
		error in vertical reading negligible	
3. Cross-hairs inclined	Section 4.3	error in both vertical and horizontal circle readings	Bisect target by same point on hair on F.L. and F.R. and take mean
4. Vertical circle index error	Section 4.4	error in vertical circle reading	Take mean of F.L. and F.R.
5. Existence of parallax	Section 4.5	error in both circle readings	Eliminate parallax
6. Optical plummet	Section 4.6	maximum error in horizontal circle reading $e\rho''/D$	Adjustment or use plumb-bob

5 Accuracies of constructional features of theodolites

This chapter is concerned with those defects in construction which it is generally not possible for the surveyor to put right, but which are reduced by design principles and by suitable observing techniques.

5.1 Trunnion axis not perpendicular to the vertical axis

Theodolites are designed so that the trunnion axis is perpendicular to the vertical axis. When a theodolite is set-up and levelled accurately (section 4.1) the trunnion axis will be horizontal within the tolerances of construction. Dislevelment of the trunnion axis (i.e. its inclination to a horizontal plane) when the vertical axis is truly vertical is caused by one end of the axis being higher than the other. This error is sometimes referred to as the *height of standards error*. Kerschner, 1980 gives a value of 5″ for the maximum angular dislevelment of the trunnion axis from this cause in the Hewlett-Packard 3820A Electronic Total Station. There will be variations of dislevelment as the telescope is rotated about the trunnion axis and these are often referred to as *axis wobble*. A maximum angular value of trunnion axis wobble quoted by Kerschner, 1980 is 1.5″ for the same instrument. Axis wobble is caused by differently shaped profiles of the ends of the axis and by lack of balance of the telescope. The effect of trunnion axis dislevelment is considered below, independent of vertical axis errors.

Suppose (Fig 5.1a) that the vertical axis is truly vertical and is along OZ, but that the trunnion axis is inclined at an angle $(90 - i_t)$ to the vertical axis. Rotation of the line of collimation about the trunnion axis results in the tracing-out of the great circle $Z'PP'_1$ instead of ZPP_1. Thus the error in azimuth is $P_1P_1' = e_t$, say. From the right-angled spherical triangle PP_1P_1' (Figs 5.1b and c),

$$\sin e_t = \tan i_t, \tan h$$

therefore

$$e_t'' \simeq i_t'' \tan h.$$

The vertical angle recorded (h') is given by

$$\sin h = \cos i_t \sin h' \text{ (Figs 5.1b and c)}$$

therefore $\sin h' - \sin h = \sin h (\sec i_t - 1)$

so $2 \sin 1/2 (h' - h) \cos 1/2 (h' + h) = \sin h (\sec i_t - 1)$.

Therefore $(h' - h) \doteqdot \sin h (1 + 1/2 i_t^2 - 1)$

and $(h' - h)'' \doteqdot 1/2 i_t''^2 \sin h / p''$

If face is changed on P, Z' is transferred to the opposite side of the zenith Z and the error e_t will be of the same size but of opposite sign. Hence the error is eliminated by taking the mean of a F.L. and a F.R. observation.

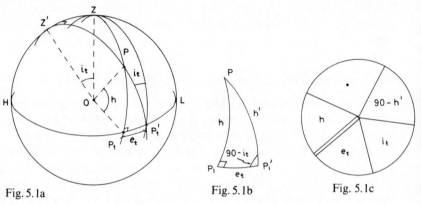

Fig. 5.1a   Fig. 5.1b   Fig. 5.1c

It should be noted that the error e_t (unlike that arising from non-verticality of the vertical axis) does not vary with azimuth.

The necessity for measuring i_t in order to correct a horizontal circle reading most often arises in astronomical observations. A striding level of higher sensitivity than the plate level is used and often requires special fitting to the theodolite. When a striding level is fitted across the transit axis it measures the algebraic sum $(i_t + i_v \sin \alpha)$ where i_v is the inclination of the vertical axis (section 4.7.1) and α is the angle measured from the plane of the axis tilt to the direction of pointing.

The development of kinematically designed hardened steel axes, lighter and shorter telescopes and the enclosure of the mounting in a casing are all factors which result in modern theodolites being virtually free of this trunnion axis error in normal use. The adjustment of the axis cannot generally be carried out by the surveyor although some manufacturers do give instructions for the adjustment (e.g. Vickers for the Tavistock II) and some make it possible to measure the inclination without a striding level (e.g. Kern with the DKM2AM). This however does not mean that there is no need to test the theodolite for this error if angles at large elevations or depressions are to be measured on one face. This test is often called the 'spire test' (section 5.1.1).

5.1.1 The spire test

A theodolite is tested for non perpendicularity of the transit axis in the following manner. The test for and adjustment of horizontal collimation error (section 4.2) should be carried out first. Then:

1. Set up the instrument so that a target at an elevation of at least 30° can be sighted. Setting up about 30 m from the face of a tall building is satisfactory.
2. Carefully level the theodolite (the altitude level can be used if it can be viewed directly and is more sensitive than the plate level).
3. On F.L., bisect a well defined target high on the building with the vertical hair. Clamp both horizontal movements.
4. Depress the telescope and note the intersection of the vertical hair with a graduated levelling staff placed horizontally at the foot of the target.
5. Repeat (3) on F.R. '
6. Depress the telescope. The line of sight should pass through the intersection noted in step (4). If not, the error is caused by dislevelment of the trunnion axis, provided the vertical axis is truly vertical.

5.2 Planes of the circles not perpendicular to their respective axes

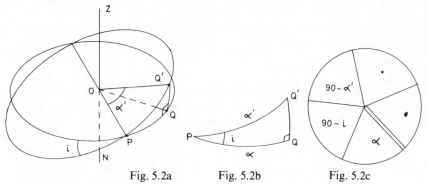

Fig. 5.2a   Fig. 5.2b   Fig. 5.2c

In Fig 5.2a
    ZON is the vertical axis.
    PQ is the correct horizontal circle (radius R).
    PQ′ is the inclined horizontal circle (i to the horizontal).

$\alpha = P\hat{O}Q$ is the angle between P and Q in a horizontal plane.
$\alpha' = P\hat{O}Q'$ is the angle α measured on the inclined circle.

Let the error in the measured angle be e,

then $$e = (\alpha' - \alpha)$$

Hence
$$\tan e = \frac{\tan \alpha' - \tan \alpha}{1 + \tan \alpha' \tan \alpha}$$

But from the right-angled spherical triangle Q'QP (Figs 5.2b and c)

$$\cos i = \tan \alpha \cot \alpha'$$

therefore
$$\tan \alpha = \tan \alpha' \cos i$$

so
$$\tan e = \frac{\tan \alpha' (1 - \cos i)}{1 + \tan^2 \alpha' \cos i}.$$

But i is small, so $\cos i = 1 - \dfrac{i^2}{2!}$ to second order,

and therefore
$$\tan e \simeq \frac{i^2 \tan \alpha'}{2 \sec^2 \alpha' - i^2 \tan^2 \alpha'}$$

so
$$\tan e \simeq \frac{i^2 \sin \alpha' \cos \alpha'}{2 - i^2 \sin^2 \alpha'}$$

Because i^2 is a quantity of the second order, $2 - i^2 \sin^2\alpha' \simeq 2$,

therefore
$$\tan e \simeq 1/2 \, i^2 \sin \alpha' \cos \alpha'$$

so
$$e'' \simeq 1/4 \, i''^2 \sin 2\alpha'/\rho''$$

and for e to amount to 1″ an obliquity of 15′ is necessary.
Errors of this magnitude ought not be present in modern instruments.
If they were, it would be apparent from other factors which would
prevent the theodolite from being used at all.

5.3 Circle eccentricities

The centre of the graduations of the horizontal circle should coincide
with the centre of rotation of the alidade. A small amount of axis
wobble will cause an eccentricity between the two centres. Similarly,
the centre of the graduations of the vertical circle should coincide with
the trunnion axis.

5.3.1 With one index

In Fig 5.3, C is the centre of graduations, C′ is the centre of rotation of
the index J, and d = CC′, the eccentricity.

Let the radius of the circle be R and assume that the alidade is
directed towards a point P at an angle α to the direction of the
eccentricity C′C. Then instead of the index being at I, it is at J and the
error in the circle reading is e. Because d is of the order of 10^{-6} m and
CP is of the order of hundreds of metres, CP can be taken as parallel to
C′P.

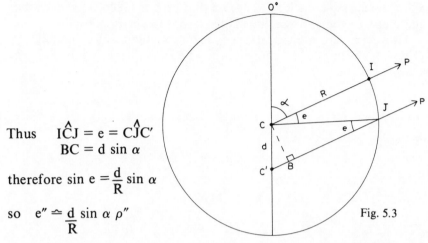

Thus $\quad I\hat{C}J = e = C\hat{J}C'$
$\quad\quad\quad BC = d\ \sin \alpha$

therefore $\sin e = \dfrac{d}{R}\ \sin \alpha$

so $\quad e'' \simeq \dfrac{d}{R}\ \sin \alpha\ \rho''$

Fig. 5.3

So e will have maximum numerical values when $\alpha = 90°$ and $270°$ and will be zero when $\alpha = 0°$ and $180°$. When $\alpha = 90°$, $e'' \simeq (d/R)\ \rho''$ and for this to amount to $1''$ on a 45 mm radius circle the eccentricity d would have to be about $0.2\ \mu$m.

Eccentricities of a few micrometers can develop in lower-order instruments so that in theodolites fitted with one index (e.g. optical scale theodolites) errors from this source can become significant. They can be eliminated by observing on two faces and taking the mean; if e_L is the error on F.L. and e_R is the error on F.R., then

$$e_L = \dfrac{d}{R}\ \sin \alpha \quad \text{and} \quad e_R = \dfrac{d}{R}\ \sin (\alpha + 180) = -e_L$$

5.3.2 With two indices

In Fig 5.4, J_1 and J_2 are two indices rotating about the eccentric centre

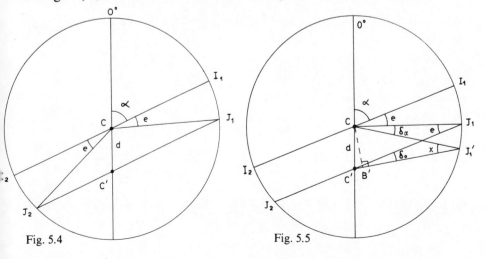

Fig. 5.4

Fig. 5.5

C'. The reading at J_1 is $(\alpha + e)$ and that at J_2 is $(180 + \alpha - e)$ so that the mean of the two readings is free from the error 'e'.

However, suppose that index J_1 is out of alignment by an angle δ_0 (Fig 5.5) and is at J'_1. There will then be an additional error δ_α in the reading at J'_1.

$$CB' = d \sin (\alpha + \delta_0)$$

Therefore
$$\sin x = \frac{d}{R} \sin (\alpha + \delta_0)$$

so
$$x \simeq \frac{d}{R} (\sin \alpha + \delta_0 \cos \alpha)$$

But
$$\delta_0 + x = \delta_\alpha + e.$$

Therefore
$$\delta_0 + \frac{d}{R}(\sin \alpha + \delta_0 \cos \alpha) \simeq \delta_\alpha + \frac{d}{R}\sin \alpha$$

so
$$\delta_\alpha \simeq \delta_0 \left(1 + \frac{d}{R}\cos \alpha\right) \quad \text{(to 1st order)}$$

This additional error δ_α will have its maximum value when $\alpha = 0$. Then

$$\delta_\alpha = \delta_0 (1 + d/R)$$

Its minimum value occurs when $\alpha = 180°$. Then

$$\delta_\alpha = \delta_0 (1 - d/R)$$

Also, $\delta_\alpha = \delta_0$ when $\alpha = 90°$ and $270°$.

Consider now the effect of the misplaced index J_1' on the measured angle α.

1. When $\alpha = 0$,   error at $J_1' = \delta_0 \left(1 + \frac{d}{R}\right)$

error at $J_2 = 0$

therefore,   error of the mean $= \frac{\delta_0}{2} \left(1 + \frac{d}{R}\right)$

2. For α,   error at $J_1' = \frac{d}{R} \sin \alpha + \delta_0 \left(1 + \frac{d}{R}\cos \alpha\right)$

error at $J_2 = \frac{-d \sin \alpha}{R}$

therefore,   error of the mean $= \frac{\delta_0}{2} \left(1 + \frac{d}{R}\cos \alpha\right)$

Thus the difference between the means will be in error by approximately

$$\frac{\delta_0}{2}\left(1 + \frac{d}{R}\cos\alpha\right) - \frac{\delta_0}{2}\left(1 + \frac{d}{R}\right) = \frac{\delta_0}{2}.\frac{d}{R}\cos\alpha \text{ with a maximum of} \frac{\delta_0}{2} \times \frac{d}{R}.$$

For example, if the misalignment of the index J_1' is $1°$, $d = 10^{-2}$ mm and $R = 45$ mm then the maximum error in an angle measured from the direction $C'C$ is

$$\frac{3600''}{2} \times \frac{10^{-2}}{45} = 0''.4$$

Therefore it can be seen that a misalignment of an index even by as much as $1°$ has very little effect on the measurement of an angle.

5.3.3 Variation of eccentricity

In sections 5.3.1 and 5.3.2 it has been assumed that the eccentricity d remains constant during the measurement. In practice, this is unlikely; the eccentricity d will vary in size and in direction owing to wobble of the axis of the alidade in its constraint. A thin film of oil between the sleeve and the axis will allow a small movement (of the order of ± 0.4 μm) about the mean film thickness of $0.5\,\mu$m for a $1''$ instrument (Haller, 1963). This will produce an angular deviation of $\pm 1.2''$ to $\pm 1.6''$ for an axis 100 mm long. The Kern system illustrated in section 3.2.1 is claimed to have a greater stability owing to the larger diameter of the ball race ($\pm 0.35''$ according to Haller, 1963).

The amount of eccentricity will then fluctuate (Fig 5.6c) between zero (Fig 5.6a) and a maximum value d (Figs 5.6b and d).

The direction of eccentricity can take up any value between $0°$ and $360°$. This error is a random error assuming that the alidade is balanced and its effects are therefore reduced by taking several observations of the angle, the alidade being rotated a few times

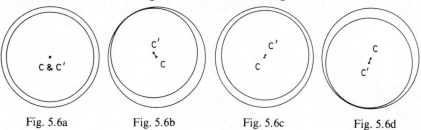

Fig. 5.6a   Fig. 5.6b   Fig. 5.6c   Fig. 5.6d

between sets. The centre of rotation of the circle may not coincide with the centre of the graduations, (the circle axis itself wobbling in its constraint) so the position of the circle should also be changed for each set of observations. In this respect, the reduction of the effects of this error is similar to that for reducing the effects of circle graduation errors (section 5.4).

A method for detecting and measuring this error using the plate level is described by Szymónski, c. 1960. The maximum inclination for a Zeiss (Jena) Theo 010 was found to be 2.5″.

5.4 Circle graduation errors

The determination of circle graduation errors is generally of more interest to the manufacturer than to the surveyor. Test methods are usually used to monitor samples from the production-line and these methods are not designed to provide sufficient information for reliable corrections to be applied to circle readings.

The errors in the graduations can be divided into two; long-period and short-period errors. The former are caused by irregularities in the spur-gear of the dividing machine and have a period 2π. However since the determination of such errors is only justified for higher-order theodolites (in which coincidence reading is used), these errors have an effective period π. Short-period errors are caused by errors in the concave screw (globoid worm) and in the cog wheel which connects it to the dividing machine.

5.4.1 Test methods

The best-known method is that of Heuvelink, 1925. This enables the long-period diametral error to be determined as a function of the position on the circle. A test angle is measured on n equally spaced settings of the circle. On the assumption that the error curve can be expressed as the Fourier series

$$e = \sum_{i=1}^{k} a_{2i} \sin (2i\phi + \omega_{2i})$$

the coefficients a_{2i} and the constants ω_{2i} can be estimated from a least squares adjustment of the observations. (In practice k is usually taken as 3).

Various test systems and analyses of results have been described by Weise, 1966 and a development using stops against the alidade to define the test angle (limit-stop testing) is described in detail. An error curve for a Zeiss (Jena) Theo 010 is shown in Fig. 5.7. A recent method which makes use of laser interferometry for calibrating horizontal circle graduations is described by Maurer, 1981.

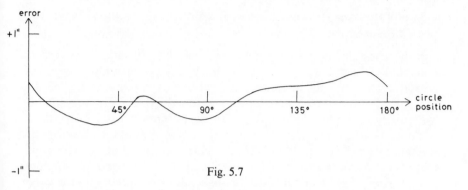

Fig. 5.7

5.4.2 Reduction of the errors

It has already been mentioned (section 5.3.3) that the effects of this error can be reduced by observing an angle on different settings of the horizontal circle, and averaging the results. The determination of the error curve entails the use of special equipment and analysis techniques. It is therefore only justified in extreme cases when other factors (notably refraction) which influence the observations can be controlled or allowed for. Setting-out precision equipment and machinery in laboratories is one problem which can be carried out more effectively if circle graduation errors are known. An example of zero settings to be used for repeated measurements of a horizontal angle is given in section 5.7.

5.5 Axis strain

This source of error was first noticed and described by Rannie and Dennis, 1934 and 1936. It arises partly from a slow release of internal stresses set up by the heat produced in high speed grinding of hardened steel axes. This in turn introduces a displacement of the optical components in the reading system.

The error is constant if the theodolite is used in the normal way but it can be detected if the position of the bearings relative to the axis is changed. Rannie and Dennis devised the following method for detecting the error:

1. Take several observations of the same angle in the usual way. Take the mean.
2. Remove the theodolite and replace it with the tribrach turned through 120°. This alters the relationship between the axis and its bearings as far as the observed angle is concerned.
3. Re-measure the angle several times and take the mean.

4. Rotate the tribrach through another 120° and repeat the measurements.

If there is a difference between the three sets of observations greater than that indicated by the spread of each set, axis strain is present. A similar effect can be introduced by misuse of the theodolite; if the horizontal clamps are persistently tightened excessively, a displacement of the members carrying the optical components will result.

To make any conclusions clearer, the angle should be measured on the same part of the circle each time to eliminate graduation errors and some system of constrained centring should be used to prevent miscentring after each rotation through 120°. The errors on the early types of instruments were several seconds in magnitude but a re-designed bearing system reduced them. On a re-designed first-order instrument, the axial strain error will probably be less than 1″ but for primary observations Rannie and Dennis recommended rotation of the tribrach and averaging of the results. This procedure was adopted and described by Watt, 1963 for the Zeiss (Oberkochen) Th 3, with the conclusion that if axis strain errors were present they were less than 1″ in magnitude.

5.6 Centring errors

The accuracies of centring with a plumb-bob, centring-rod and optical plummet have been given in section 2.6 as about 1 mm.

When considering the accuracies of constrained centring, the relevant criterion is the repeatability of the system and not the absolute accuracy of centring over a ground mark. In other words, one is concerned with the degree to which the vertical axis of a theodolite placed in a tribrach coincides with the vertical through the centre of the target previously occupying the tribrach.

It is convenient to consider any tribrach as defining a 'centring axis' and to see to what extent the vertical axis of a theodolite or the axis of rotation of a target coincides with this centring axis.

For a self-locating system using the three-stud method of constraint (e.g. Wild – see section 3.6) the tribrach slots at 120° separation define the centring axis. A self-locating system using a conical stub (e.g. Kern) will have a centring axis defined by the bush in the base unit. A system using a stub axis and lateral clamp has a centring axis defined by the position of the vertical axis of a stub of the nominal diameter of the system (e.g. 34 mm).

Thus, on any given tribrach with a centring axis C defined above, a theodolite or target with vertical axis V will result, in practice, in systematic eccentricity e_s between C and V.

Berthon Jones, 1963 has investigated the magnitude of this eccentricity for different target systems by rotating V about C and using a travelling microscope to measure the displacements of the target. This rotation in the case of the Wild system for example was carried out by placing a given stud in each of the three tribrach slots in turn. V thus described a circle radius 'e_s' about the centring axis C. Some results of determination of e_s are given in the table on page 158.

In addition to this systematic error, there will be random errors (e_r) present which will cause variations in e_s for replacements in a given orientation. For example in the Wild system, replacing a given stud in a given slot will give variations in e_s. In general, these variations will arise for the following two reasons.

1. Differences in clamping pressures. This applies particularly to the stub-axis and lateral stop system. The clamp should be tightened just sufficiently to prevent rotation of the stub. If excessive pressure on the stub is introduced by the clamp, the stresses set up could relax in time thereby causing gradual miscentring during observations. In the case of the self-locating systems, the clamping pressure is determined by the manufacturer and applied by means of a spring loaded clamp.
2. Bad fits. This is a source of error which affects those systems which depend on close contact between mating surfaces, i.e. the stub axis and lateral stop systems. Dirt or oil on the surfaces and damaged components will cause random errors, and in any case there will be a certain tolerance in manufacture.

    In addition to the above sources of random errors which occur for both theodolite and target constraint, two further sources are relevant to the constraint of targets.

3. Tilt of the axis of rotation in the plane of the target. This will result in an eccentricity of the target with respect to the centring axis.
4. Eccentricity arising from re-levelling. When a target is placed on a tribrach previously occupied by a theodolite, slight re-levelling is often necessary. If the axes of the footscrews are not parallel to the centring axis, the target will be eccentric to the centring axis. However, this error will be very small in all cases except those involving footscrew replacement.

Some results of the determination of e_s and e_r are given in the following table.

Type	e_s (mm)	e_r (mm)	e_t (mm)
Cylindrical stub	0.025	0.007	(Theodolite)
Lateral stop	0.136	0.004	0.159
	0.021	0.004	(Theodolite)
Conical stub	0.123	0.010	0.031
Lateral stop	0.049	0.007	(Theodolite)
	0.022	0.016	0.480
Self-locating	0.020	0.006	(Theodolite)
Conical stub	0.005	0.013	0.147
3-stud replacement	0.006	0.022	0.077
Footscrew replacement	0.109	0.015	0.177
	0.090	0.017	(Theodolite)

In addition, the eccentricity of the centre of the target to the axis of rotation (V) was investigated by Berthon Jones, 1963 and some results are shown as e_t in the above table.

5.6.1 Reduction of centring errors in traversing

It has been shown (section 4.8) that the error in a pointing arising from an eccentricity e perpendicular to the direction of sight is approximately

$$\alpha'' = e\rho''/D$$

where D is the distance sighted.

Thus for a traverse angle ABC, the observed angle will be in error by

approximately $2\,e\rho''/D$

if $AB = BC = D$ and angle $ABC = 180°$.

This can be treated as a standard error arising from angular observations in a straight traverse of n equal legs. To get some idea of the accuracy of centring required for a given allowable bearing misclosure, consider a closed straight traverse of n legs. There will therefore be (n + 1) angles observed and the standard error of the closing bearing arising from random centring errors alone will be

$$\sigma_M'' = \pm\, 2e\rho''\, (n + 1)^{1/2}/D$$

Thus $e = \pm\, \sigma_M''\, D/2\rho''\, (n + 1)^{1/2}$

For a 1 km traverse, necessary values of e (mm) for a given standard error of the final bearing σ_M'' are as follows:

	D(m)		
σ_M''	100	25	10
60	4.38 mm	1.09 mm	0.44 mm
30	2.19	0.54	0.22
10	0.73	0.18	0.07

It must be pointed out that the above figures should be treated with caution; for a start the bearing misclosure will in practice be affected by other random errors (in pointing and reading the circle for example) and of course by systematic errors. In fact it is likely that the effects of miscentring will have a systematic influence since at each station the tripod will probably be set in the same attitude relative to the adjacent traverse lines to enable the observer to make the sightings comfortably without having to straddle the tripod legs. Thus the tribrach will not be placed in random positions. For these reasons, the actual allowable centring error for a given standard error of the final bearing will be smaller than the figures given above.

It can be seen that the equipment tested by Berthon Jones, 1963 will not always give the required accuracy. Methods of reducing the effects of constrained centring errors are therefore worthy of consideration.

Allan, 1977 discusses the limitations of centring using a spherical level in conjunction with EDM instruments.

5.6.2 Reduction of centring errors by field techniques

The usual procedure in three-tripod traversing is to observe the angle at the nth station between target A at the (n − 1)th station and target B at the (n + 1)th station. Then target B is moved to the (n + 2)th station the theodolite to the (n + 1)th station and target A to the nth station etc.

A better procedure is described by Berthon Jones, 1963 whereby after observing the nth angle, target A is placed at the (n + 2)th station and target B at the nth station, thus each target is used alternately as a backsight.

It is shown that this latter procedure gives an improvement of n time in σ_M'' where n is the number of traverse legs. Such a procedure will take longer than the usual method, but since it is only relevant when the legs are short, not much extra carrying will be necessary.

In order to take account of the eccentricity of the theodolite, the obvious method is to rotate the instrument in its tribrach, measure the angle on three settings at 120° separation and take the mean. Remembering that these techniques are relevant only when high accuracy is aimed at it will be seen that this procedure will not slow down work too much since the angle could well have to be read more than three times anyway to reduce random errors of reading and pointing.

5.6.3 Reduction of centring errors by special equipment

Tarczy-Hornoch, 1967 introduced the MOM system for eliminating e_s eccentricity by rotating the sleeve housing the stub axle through 180° about the centring axis. This is more convenient in some ways than rotating the stub axis itself in its sleeve since the precision-ground mating surfaces are not thereby exposed during orientation. This was then developed for use with reiteration type theodolites whereby the sleeve housing the stub axis could be rotated by means of a slow-motion screw, thus enabling the theodolite to be used as a repetition theodolite.

Tests carried out with this system by the manufacturers using lines as short as 1.5 m gave no indication of residual eccentricities greater than 0.01 mm.

5.7 Errors in the optical micrometer

Ideally, rotation of the optical micrometer from its zero scale graduation to its maximum (e.g. 10′) scale graduation results in the image of a graduation on the main scale being displaced by exactly the same amount relative to the index. Mechanical and optical defects in the micrometer will be present, however, and the ideal case will never occur in practice.

5.7.1 Micrometer run-error

If the magnifications of the images of the micrometer scale and circle are incorrect, rotation of the micrometer through its nominal range r will cause the image of a circle graduation to be displaced apparently by an amount $r + \delta r$ relative to the index. The small error δr will be positive if the magnification of the circle is too small relative to the magnification of the micrometer scale and it will be negative if the magnification of the circle is too large.

The error δr is known as the micrometer run-error. It is assumed that the error in magnification is constant over the range of the micrometer, so δr is a constant for a given instrument. The traditional method for the determination of the run-error is as follows:

1.   Set the micrometer near that end of its run corresponding to the maximum reading so that one circle graduation is against the

index. Record the micrometer reading (r_1) and the main scale reading (R_1).

2. Set the micrometer so that the next highest circle graduation is against the index. Record the micrometer reading (r_2) and the main scale reading (R_2).

3. The nominal run of the micrometer is $r = R_2 - R_1$ and the micrometer run-error is

$$\delta r = (R_1 + r_1) - (R_2 + r_2)$$

4. Repeat steps 1–3 and obtain a series of values of δr from which a mean value can be found.

When this method is applied to modern optical micrometer theodolites, it is necessary to take the readings r_1 and r_2 close to the ends of the micrometer scale (zero and r marks) and only when the micrometer graduations extend before the zero and beyond the r graduation is it possible to apply this method successfully. In the example illustrated in Fig 5.8a, the nominal range r of the micrometer is 20′. The readings are: $R_1 = 3°\ 20′$ and $r_1 = 20′\ 03″$, so that the full reading is $(R_1 + r_1) = 3°\ 40′\ 03″$. For the same telescope pointing, Fig 5.8b illustrates readings $R_2 = 3°\ 40′$ and $r_2 = -1″$, so that the full reading $(R_2 + r_2)$ is $3°\ 39′\ 59″$. Therefore the micrometer run-error is:

$$\delta r = 3°\ 40′\ 03″ - 3°\ 39′\ 59″$$
$$= +4″$$

This indicates that the micrometer scale has been used to measure the 20′ interval of the main scale and has given 20′ 04″ as the result, so the magnification of the main scale is relatively too small.

If it is assumed that the error has arisen from a uniformly excessive magnification of the image of the micrometer scale relative to the image of the circle, then at any other micrometer reading r_i, the error is $\delta r_i = r_i . \delta r/r$ and if δr is known, a correction $-\delta r_i$ can be applied to the micrometer reading r_i.

In modern theodolites, the micrometer run-error is likely to be no more than a few seconds in magnitude and is often less, but if the highest accuracy of measurement is aimed at, then the error should be determined and measurements subsequently corrected if necessary. This is generally done in workshop or laboratory surveying where other sources of systematic error can be either controlled or calibrated.

If the determination of the micrometer run-error is to be regarded as free from the effects of circle graduation errors, then several determinations of the error should be made over different parts of the circle circumference.

If a theodolite does not have graduations outside the range r of the

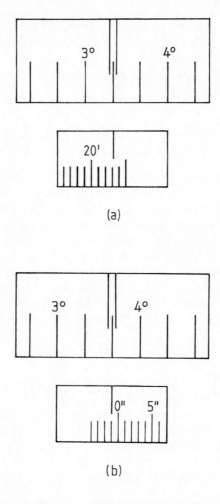

Fig. 5.8

micrometer scale, the micrometer reading will have to be estimated if the index mark falls outside the range for one or more settings.

It is necessary to carry out separate determinations of micrometer run-error for horizontal and vertical circle readings because the magnifications of the two circles may differ.

5.7.2 Micrometer cyclic errors

These are likely to arise from cyclic errors in the mechanical gearing of the micrometer.

If these latter defects are to be investigated it is necessary to carry out the work so that the circle graduation errors cannot interfere with the micrometer errors. Thus if a small angle is measured several times on the same setting of the horizontal circle but with different micrometer settings, deductions about the micrometer errors can be made.

Dyer, 1958 used a procedure for investigating these errors in an (unspecified) 1″ theodolite and found that they exhibited a cyclic variation of period 10′ over the 10′ run of the micrometer scale, with a total range of about 10″ between maximum and minimum. The most likely source of this variation is in the mechanical linkage which connects the micrometer (wedge or parallel plate) to the micrometer scale. The effects of this error can be reduced by measuring an angle on different settings of the micrometer. This is coupled with the procedure for reducing circle graduation errors by observing on different settings of the horizontal circle.

For example, if the coincidence method is used, the period of the errors of the graduations on the horizontal circle is 180° and the settings for n rounds should be distributed evenly between 0° and 180°. If the run of the micrometer is 10′, this is the period of the micrometer errors and the micrometer settings would be distributed evenly between 0′ and 10′.

An example of suitable zero settings for 16 rounds of angles is given below

Round	Setting			Round	Setting		
1	00°	01′	05″	9	11°	16′	05″
2	90	08	55	10	101	23	55
3	45	02	10	11	56	17	10
4	135	07	50	12	146	22	50
5	22	33	20	13	33	48	20
6	112	36	40	14	123	51	40
7	67	34	30	15	78	49	30
8	157	35	30	16	168	50	30

5.8 Accuracies of automatic vertical circle indices

One method of testing the indices is described by Folloni, 1965 for the Salmoiraghi 4200. Basically, the method consists of taking several observations of the vertical angle to a fixed target, the vertical axis of the theodolite being moved out of the vertical between each measurement. This movement is made using a footscrew and is enough to take the compensator out of its working range and back again. The resulting spread of readings has to be cleared of the effects of sighting and reading errors before drawing any conclusions. The results indicate that the standard error of the index is ± 0″.21.

Most manufacturers claim a stabilisation to ± 1″ approximately on lower order (20″) theodolites and Kern quote ± 0″.3 for the stabiliser on the 1″ DKM 2-A.

5.9 Summary and hints for observers

It will not generally be necessary to use the special techniques or corrections described in this chapter. However, for each surveying task, a certain accuracy should be aimed at and this will determine whether or not the above sources of error are significant. If they are, then either calibration should be carried out, or a suitable technique of observing must be derived and followed, in the expectation that the errors will thereby be reduced. If any doubt exists, then suitable tests or checks must be carried out to see if the required accuracy has been achieved. This should be normal survey practice.

The table opposite summarises the errors, their effects and appropriate actions discussed in this chapter.

It has been seen that the elimination or reduction of the effects of instrumental maladjustments and malconstructions on observations can in many cases be achieved by suitable observing procedures. The following general points should also be borne in mind by the observer.

1. The more quickly the observations are made, the more likely it is that consistent results will be obtained. Differential heating of the wooden tripod causes the theodolite gradually to go off-level and off-centre. The tripod will slowly settle or rise (soil rebound effects are discussed in section 9.6.3) causing additional levelling and centring errors. If a round of angles takes a long time to complete, a significant time-dependent systematic error will probably be introduced into the sequence of readings.

2. The theodolite should never be re-levelled or re-centred during observations, but left until the round is completed. Levelling and centring should always be checked before starting a round of observations.

3. There is no need to tighten clamps more than is necessary to prevent accidental movement of components. Over-tightening can lead to strains in the optical and mechanical components which will introduce inaccuracies.

4. The theodolite should be used with a light touch and the tripod should not be touched by the observer during observations. Some thought during setting-up could ensure that the tripod legs are positioned in relation to the targets so that it is not necessary to straddle a tripod leg during observations.

5. Final setting of slow-motion drives for the alidade, micrometer, telescope, vertical circle index level and so on, should be made by a clockwise rotation to move the driven unit. This will reduce the effects of backlash in the slow-motion screw thread.

Source of error	Section	Effect	Action taken
1. Trunnion axis not perpendicular to vertical axis (Error = i_t)	5.1	(a) Error in horizontal circle reading $e_t'' \fallingdotseq i_t'' \tan h$ (b) Error in vertical circle reading $(h' - h)'' \fallingdotseq 1/2\, i_t''^2 \sin h/\rho''$	(a) Observe on F.L. and on F.R. and take the mean (b) Error negligible in practical cases
2. Plane of circle inclined to its axis. (Error = i)	5.2	Error $e'' \fallingdotseq 1/4\, i''^2 \sin 2\,\alpha/\rho''$	Error negligible in practical cases
3. Centre of circle graduations eccentric to axis of rotation (Error = d)	5.3	(a) With one index, error $e'' = (d/R)\sin\alpha\ \rho''$ (b) With two indices, one misaligned δ_0	(a) Observe on F.L. and F.R. and take mean (b) Additional error negligible
4. Inclination of alidade axis in its sleeve.	5.3.3	Inclination of vertical axis a few seconds.	Rotate the alidade between rounds.
5. Circle graduation errors	5.4	Errors of a few seconds in readings	(i) Observe on different settings, or (ii) calibrate
6. Axis strain	5.5	Errors of about one second in readings	Rotate tribrach on tripod or pillar between rounds.
7. Constrained centring errors	5.6	Eccentricities of the order of 0.1 mm	(i) Rotate the measuring unit within the constraint between rounds and take the mean (ii) Suitable field procedure (section 5.6.2) (iii) Use special equipment (section 5.6.3)
8. Optical micrometer errors	5.7	Errors of a few seconds	(i) calibrate to determine micrometer run-error and apply corrections, and (ii) observe with different initial micrometer settings
9. Automatic vertical circle index error	5.8	Error of about 1″	Use footscrew to 'swing' stabiliser within its workings range between readings and take mean

6. For consistent results, horizontal angles and vertical angles should be observed separately. If possible, vertical angles over distances greater than about 1 km should be measured in the early afternoon, say noon until 2 pm, when atmospheric conditions are likely to be most stable, and correction terms for refraction most reliable.

7. The theodolite should be set up at a height convenient for the observer. Fatigue soon sets in if the instrument is too high or too low and accuracy will deteriorate.

8. If the observer has somebody to act as a booker, then the former should ensure that the booker checks the readings as they are made for gross errors, rather than wait until the end of a round. If a gross error is made, say in reading the minutes of a coincidence micrometer, it is easier to re-check this before changing the pointing. Of course, it may be that the original reading was wrong and the current one correct so care should be exercised in amending readings. If in doubt, the round should be cancelled and repeated.

9. It is good practice to spend a few minutes checking the readings and reductions before taking down the theodolite and tripod. It is often difficult and expensive to have to return to a station and re-observe.

10. When horizontal angles are to be measured between more than two stations, the station which is easiest to bisect accurately should be chosen as the reference object (or RO) and rounds of angles should be started and finished on that station.

11. The targets should be chosen so that when sighted through the telescope, they appear no more than slightly larger than the cross-hairs. This makes accurate bisection easier than with a larger target image.

12. For precise measurements, when on face-left the unclamped alidade should be swung so that the objective of the telescope moves to the right (i.e. clockwise when viewed from above) and *vice versa*. This tends to reduce sudden changes in the vertical axis wobble.

Although in the foregoing, consistency of data has been said to be desirable, it should be remembered that measurements can be consistently wrong. For instance, the centring can be in error by tens of millimetres and this will not affect the consistency of the observations. Inconsistent data should be used to detect and eliminate systematic errors and blunders.

The method of booking or automatically recording the observations is partly a matter of personal preference but is governed largely by the observing procedure. This in turn depends upon the accuracy required, the layout of the stations and the way in which the observations are to be used in computations. In addition, there may be some legal requirements to be met. In view of these factors, it is not possible to lay down a universal method for making and recording observations, but the following three examples are in common use.

In Fig. 5.9 an example is given which shows the order in which the observations of a horizontal angle of a traverse can be made and recorded. Fig 5.10 illustrates the method with a numerical example.

At	To	Face	Horizontal circle	Horizontal angle	Mean angle
C	B	L	1		
	D	L	2	$5 = 2 - 1$	$7 = (5 + 6)/2$
	+				
	D	R	3	$6 = 3 - 4$	
	B	R	4		

Fig. 5.9

At	To	Face	Horizontal circle	Horizontal angle	Mean angle
C	B	L	46° 28′ 19″		
	D	L	231 12 47	184° 44′ 28″	184° 44′ 31″
	+				
	D	R	169 37 52	184 44 34	
	B	R	344 53 18		

Fig. 5.10

It is assumed that the traverse runs from B through C to D. The first pointing is backwards to B and the second is forwards to D. The + sign indicates that the horizontal circle (or bottom plate) is deliberately rotated between the second and third readings. The reason for this change of zero in normal traversing is not so much to reduce the effects of graduation errors, but more importantly to provide an independent check on the first two readings. If the zero is not changed between the second and third readings, the observer could be influenced in taking the third reading (to D) by the recently recorded and probably remembered second reading (also to D). In such a case, the third reading would not be independent, hence the desirability of changing zero.

The horizontal angle is deduced by subtracting the first and fourth readings *from* the second and third respectively. If this is done, the resultant value is always that of the angle measured clockwise from the rear station to the forward station. If reading 2 is less than reading 1 (or reading 3 less than reading 4) the subtraction is made by first adding 360° to the smaller value. If this procedure is followed throughout the traverse, there should never be any doubt as to whether the angle appearing as the mean is that measured clockwise or anti-clockwise.

The clockwise angle is the one used at each station to compute the bearings through the traverse from one line to the next.

An observing procedure when more than two stations are to be sighted is slightly different and is illustrated in Fig 5.11. The selected

At	To	Horizontal circle						Horizontal
		Face left	corr.	Face right	corr.	Mean		angle
X	P	0° 20′ 23″	0	180° 20′ 25″	+3	0° 20′ 25.5″		
	Z	62 38 44	−1	242 38 50	+2	62 38 47.5		62° 18′ 22.0″
	R	109 52 12	−3	289 52 13	+1	109 52 11.5		109 31 46.0
	T	268 14 57	−5	88 15 07	0	268 14 59.5		267 54 34.0
	P	0 20 30	−7	180 20 28	0	0 20 25.5		0 00 00.0

Fig. 5.11

RO is P and readings are taken around the circle, clockwise on face-left, closing on the RO with a +7″ error which is distributed evenly through the round. The assumption behind this is that the error has arisen uniformly as described in general point 1 above. If there is any evidence to the contrary, then another appropriate method of dealing with the +7″ misclosure should be used.

After closing the horizon on face-left, the RO is sighted again but this time on face-right and the stations are then sighted in reverse order (P,T,R,Z), finally closing on P with an error of −3″. This error is treated similarly to the distribution of the +7″ misclosure on face-left. The mean 'corrected' readings of minutes and seconds are derived and given the degrees of the face-left readings. The horizontal angles are then deduced with reference to the RO.

The foregoing example constitutes one round on one zero. At least one other round in a different zero is needed for an independent check against gross errors. It may not be necessary (or even advisable) to 'correct' and reduce the angles in the way shown for further computation. It will depend, for example, on whether angles or uncorrelated directions are required. But the foregoing method of booking does give suitable checks against gross errors in the field. Directions or different angles can be deduced from the recorded data if they are needed for computations.

Vertical angles are generally not recorded in rounds but in pairs; a face-left and face-right reading are made successively to the same station. A field method for recording and reducing the vertical angles to the stations of the foregoing example is illustrated in Fig 5.12. There appears to be a collimation error of about 1′ in the vertical circle index, because the face-left and face-right directions to a point show a more or less constant difference of 2′.

At	To	Face	Vert. circle	Vert. angle	Mean	t
X	P	L	89° 28′ 30″	+0° 31′ 30″		
i = 1.381 m		R	270 29 20	+0 29 20	+0° 30′ 25″	1.252 m
	Z	L	90 38 10	−0 38 10		
		R	269 19 42	−0 40 18	−0 39 14	0.284 m
	R	L	89 59 58	+0 00 02		
		R	269 57 48	−0 02 12	−0 01 05	1.505 m
	T	L	88 12 20	+1 47 40		
		R	271 45 27	+1 45 27	+1 46 34	0.207 m

Fig. 5.12

The instrument height (i) and target heights (t) are an important feature of the data to be measured and recorded for subsequent computation of height differences, but they are often forgotten by beginners.

6 Basic constructional features of the level

In chapter 1, it is stated that a convenient method of defining the positions of points is in terms of X, Y and Z rectangular co-ordinates, where the XY plane is a horizontal plane. A level is an instrument for measuring the difference between the Z co-ordinates of any two points. It consists essentially of a telescope similar to that used in theodolites, where the *line of collimation* (see section 1.1) is adjusted to be horizontal.

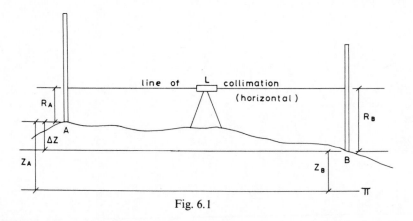

Fig. 6.1

Fig 6.1 shows two points A and B with Z co-ordinates Z_A and Z_B respectively, referred to the horizontal (XY) datum plane π. If a level L is set up and pointed to a staff graduated in the required units from zero at its foot and held vertically on A, a staff reading (R_A) is obtained. This reading is the intersection of the horizontal line of collimation with the staff graduations. If the staff is transferred to B and held vertically there, a reading R_B is obtained. The difference in the readings ($\triangle R$) is equal to the difference between the Z co-ordinates ($\triangle Z$) of A and B. More fully, with regard to signs,

$$(Z_A - Z_B) = (R_B - R_A)$$

This process of reading a staff in order to determine ΔZ is called *levelling*. The beginner should not infer from Fig 6.1 that levelling can only be carried out with the level set up on the line between successive staff positions A and B. It is normal to illustrate diagrammatically the practice of levelling by projecting the level and staff positions onto a vertical plane and this gives the impression that the level and consecutive staff positions are collinear. For reasons which are discussed later in this section and in chapters 9 and 10, it is good practice to ensure that the level is equidistant from successive staff positions, but it is not necessary to place the level midway between them to achieve this.

The plane π is a *datum surface*, and the Z co-ordinates of points such as A and B are generally called heights, or *reduced levels*. (R.L.s). In practice, the datum surface to which reduced levels are referred is not a plane. It is a slightly irregular surface (almost a spheroid) which is everywhere perpendicular to the direction of gravity, or it can be described as a surface on which the gravitational potential is constant (i.e. an equipotential surface). Over small distances it can be considered as a plane; the datum surface deviates from a plane by an amount of the order of 1 mm over a distance of 100 m.

There are an infinite number of equipotential surfaces, so if one of these is to be used as a datum, it has to be further defined. In Britain, the datum surface is that equipotential surface which passes through the mean level of the sea determined by observations at Newlyn in Cornwall and this is Ordnance Datum (O.D.). Similar definitions of the datum surface are used for other countries.

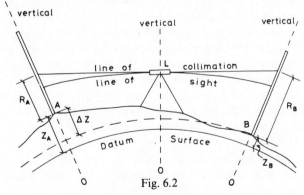

Fig. 6.2

Fig 6.2 is a modification of Fig 6.1 taking account of the curvature of the datum surface. If L is equidistant from A and B (and provided the datum surface approximates to a sphere over the range AB) then as before

$$(Z_A - Z_B) = (R_B - R_A)$$

In practice, the line of sight will deviate from the horizontal line of collimation because of refraction through the atmosphere, and readings obtained will generally be less than R_A and R_B. But again, if L is equidistant from A and B, each reading will be in error by approximately the same amount and the difference between the readings will still be correct, or nearly so.

The fact that the level is equidistant from A and B not only eliminates the effects of curvature and reduces considerably those of refraction but it also eliminates residual instrumental errors in most cases and is therefore highly desirable even in lower-order levelling.

The requirement of a level therefore is to provide a horizontal line of collimation. The earliest levels met this requirement in two different ways. Firstly, the horizontal direction was defined by the surface of a liquid, usually water, at rest in a trough. The line of sight was the surface of the water. The surveyor placed his eye at the level of the water and looked along the water surface towards the target . Instruments of this type were used by the Roman surveyor Vitruvius around 15 B.C. and were known as the *chorobates*. They were in general use for levelling in Britain until the middle of the seventeenth century.

The second method of defining a horizontal direction was by means of a sighting rule held perpendicular to a plumb-bob string. In its simplest form, the sighting rule was one side of a triangular frame, the other two sides being equal in length. A plumb-bob was suspended from the centre of the sighting rule which was tilted until the string passed through the apex of the inverted triangle. Later versions used the diameter of a semi-circle as the sighting rule, with the plumb-bob suspended so that its string intersected the graduated semi-circular arc.

Levels in both forms had been used in civilisations earlier than the Graeco-Roman for irrigation and possibly other engineering and land surveying activities, but the Romans began the systematic improvement of the instruments and developed new methods of applying them.

Apart from the telescope, the only important development in instrumental levels before the present century was the spirit level. This is attributed by Richeson, 1966 to Melchisédech Thevénot in 1666.

Modern levels still depend for their operation either on the surface of a liquid at rest (dumpy, tilting and some automatic levels) or on the effects of gravity on a freely suspended device (some automatic levels).

6.1 The dumpy level

The distinguishing feature is the fact that the telescope is rigidly attached to an axis which rotates in a sleeve. The sleeve is part of a tribrach. The axis of rotation is made vertical by means of a spirit level

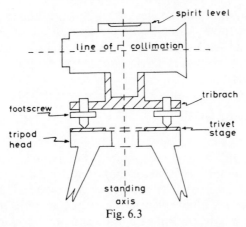

Fig. 6.3

and three footscrews and because the line of collimation should be perpendicular to this axis of rotation, it is thereby made horizontal (see Fig 6.3).

The axis about which the telescope rotates must be vertical and is therefore similar to the vertical axis of the theodolite. Tilting levels and automatic levels have telescopes which can be rotated about an axis, but for these two types of level, this axis need not be vertical, only nearly so. It would therefore be incorrect to call this axis 'the vertical axis' for tilting and automatic levels, and the term most often used is *standing axis*. It is intended to use the same term for the sake of consistency when the dumpy level is referred to. Thus for the dumpy level, the standing axis must be vertical. If the line of collimation is then perpendicular to this axis, it will trace out a horizontal plane when the telescope is rotated about the standing axis. This plane is often called the *plane of collimation*.

This type of level is obolescent and a very few manufacturers now produce more than one model of a dumpy level. If high precision is required, the stability of the standing axis must be reliable and accurate. This requirement increases the size and cost. Nearly all modern levels are either of the tilting type, or are automatic.

6.2 The tilting level

Unlike the dumpy level, which, once set up correctly, provides a horizontal line of collimation for all pointings of the telescope, the tilting level is adjusted at each pointing to give a horizontal line of collimation. This is achieved by means of a *tilting screw* which tilts the telescope about a horizontal axis at right angles to the longitudinal axis of the telescope. There is some provision for setting the standing axis approximately vertical. This can be for example, the three footscrews, or a ball and socket fitting (*quickset head*) for attaching the level to the tripod. A small spherical spirit level is used to carry out this approximate setting (see Fig 6.4). Subsequently, accurate levelling of the line of collimation is carried out using the tilting screw and the main spirit level.

6.3 The automatic level

The standing axis is set approximately vertical by means of a small spherical level and either three footscrews or a quickset levelling head. The line of collimation is made horizontal by a device inside the telescope which automatically compensates for the residual inclination. This *compensator* is often essentially a suspended reflector or refractor which deviates a light ray entering the telescope horizontally through the optical centre of the objective, onto the *reticule* at the intersection of the *cross-hairs* (see section 1.1). Thus, provided the standing axis is within several minutes of the vertical (this is the working range of the compensator) the line of collimation is automatically corrected (see Fig 6.5).

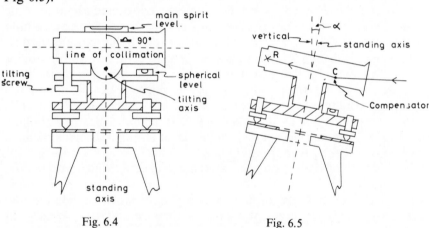

Fig. 6.4   Fig. 6.5

There is thus no need for the usual tubular spirit level. The compensation is effective for all pointings of the telescope provided the inclination of the standing axis in any direction does not exceed the working range of the compensator.

6.4 Electronic levels

When compared with the theodolite and its three axes, the level is a simpler device. It is required essentially to give a horizontal line of sight in all directions from an instrumental position and this requirement is relatively easy to meet by using optomechanical components. Automatic levels have resulted in much faster field procedures than were possible with older tilting and dumpy levels. Transformation of staff readings to give levels is relatively simple. For these reasons, the incorporation of electronic measuring and recording devices into levels has not been seen to offer the advantages that their incorporation into theodolites has offered.

On construction sites however, particularly in setting-out, the use of a laser beam to define a direction is becoming standard practice and the rotating laser level has been found to be useful for certain setting-out tasks. The principle is simple. A collimated laser beam is made to rotate about, and be normal to, a vertical axis. It therefore defines a horizontal plane which can be detected either optically or opto-mechanically. It can be set up and left whilst the operator moves around the area swept out by the beam, taking staff readings or setting-out pegs or formwork.

Examples of rotating laser levels in current production are given in section 8.5. Some of the errors in a rotating laser level are discussed in section 10.9.

7 Principles of level construction

7.1 The telescope

Level telescopes are generally the same as those on theodolites described in sections 2.1 etc. Most level telescopes are fitted with stadia lines, are internally focusing and have near-anallactic properties over normal sighting distances. Diameter of objective, magnification and resolving power vary of course depending on the precision the level is designed to achieve. On the smaller dumpy and tilting levels, the telescope is smaller than that on lower-order theodolites. Thus the resolution and magnification are low with the result that only relatively short sights should be used. Some telescopes have a combined coarse and fine main focusing screw, and on most automatic levels, the compensating system inverts the normal telescope image thus giving a final upright image.

7.2 Standing axis systems

Only the dumpy level has a requirement for the standing axis to be truly vertical. For the majority of modern non-automatic levels, the standing axis need be vertical only to within a degree or so. Thus the standing axis system need not be as accurate as that in theodolites for example where consistent verticality to the order of 1″ is often required. The standing axis system is generally cylindrical, often without a precision ball race. The weight which has to be supported by the axis system is relatively small, wear is reduced and the stability maintained.

7.3 Movement controls

The telescope can be rotated about the standing axis by a slow-motion (or tangent) screw. This screw is often associated with a clamp which connects the telescope spindle to the tribrach sleeve, the slow-motion screw being effective only when the clamp is on. When the clamp is off, the telescope can be rotated freely about the standing axis. An alternative arrangement is a device whereby part of the sleeve in which the telescope spindle fits is pressed against the spindle by a spring. The friction between the two parts enables the telescope to remain in a set direction. Slow rotation is achieved in either direction by means of a slow-motion worm screw engaging a mesh attached to the telescope spindle. Large rotation can be achieved by gripping the telescope and turning it to the required direction, by exerting enough torque to overcome the friction between the spindle and the sleeve. On automatic levels, footscrews often have large-pitch threads for rapid approximate levelling.

7.4 The horizontal circle

Lower-order levels intended mainly for use on construction sites are often fitted with a horizontal circle so that the level can be used to set-out approximate angles (e.g. for a level grid) or to enable the level to be used as a tacheometer. This graduated circle is generally rigidly attached to the tribrach, and read directly, a small portion of it being visible through a window with an engraved index below the telescope eyepiece. Some levels have a circle reading microscope and a secondary scale for estimation of fractions of the main scale graduations (normally 1°).

The manufacturers do not expect the user to use such a level as an alternative to the theodolite. However, a level with a horizontal circle does have advantages over simpler pieces of equipment such as a prism square. The circle will probably not be horizontal in use, graduation errors and eccentricities are present and accuracies better than a few minutes cannot be expected.

7.5 The spirit level

Design and construction of spirit levels are described in section 2.5. The bubble vial in its holder is mounted on the telescope. As far as levels are concerned, a dumpy level must have an exposed level vial so that the setting-up can be carried out easily. The bubble viewing system is generally an inclined plane mirror (section 2.5.3.1). A typical angular value of a 2 mm graduation is 40″. The bubble can be centred to about 0.4 mm. This means that the line of collimation can be set within about ± 8″ of the horizontal.

Most tilting levels have a prism bubble-reader (section 2.5.3.2) with the reading microscope adjacent to the telescope eyepiece. By means of this system, a bubble of the same sensitivity as that described above (i.e. 40″ per 2 mm) can be centred to about ten times the accuracy obtainable with an inclined mirror reader. Thus the line of collimation can be set horizontally to about ± 1″.

Spherical levels are used in tilting and automatic levels to set the line of collimation approximately horizontal. They are mounted on the tribrach, or on the upper part of the quickset component or on the member holding the tilting screw. A typical sensitivity for a spherical level on a tilting level is 8′ per 2 mm. Thus the initial setting can be carried out to about ± 30″, with care.

7.6 Centring

Only if a level is being used for simple setting-out or horizontal angle measurement is it necessary to set it up so that the centre of the horizontal circle graduations is vertically above some ground mark. Thus lower-order levels are often provided with a suspension point for a plumb-bob.

7.7 Reversible levels

For a tilting level to be in adjustment, the principal tangent of the spirit level vial must be parallel to the line of collimation (section 9.1). A reversible level is one in which the telescope can be rotated in V-shaped bearings through 180° about its longitudinal axis. This enables any inclination in a vertical plane between the principal tangent and the line of collimation to be detected and its effects eliminated (section 9.3).

If the spirit level is on the left of the telescope when viewed from the eyepiece end of the telescope, the level is in the 'bubble left' attitude. (Fig 7.1a) After rotation of 180° about the longitudinal axis of the telescope, the level is in the 'bubble right' attitude. (Fig 7.1b)

In use, the telescope is clamped in one position (generally bubble left). In Fig 7.1b the bubble is viewed through the liquid with the result that the image is slightly misty.

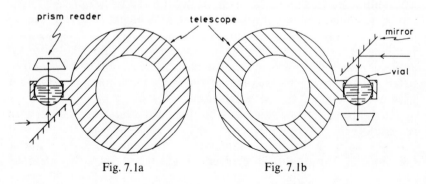

Fig. 7.1a   Fig. 7.1b

This system has the advantage that the adjustment can be checked very easily. If the surveyor has reason to believe during observations that the level has gone out of adjustment, it is possible to check this within the space of a few seconds. Otherwise the two-peg procedure (section 9.1) has to be followed. This can take several minutes and involves an interruption of the series levelling.

7.8 Quickset head

This is a device enabling the level to be fixed to the tripod with its standing axis almost vertical and its line of collimation almost horizontal. It is not present on dumpy levels (which always have footscrews) but is often present on tilting levels and lower-order automatic levels. In its usual form, it consists of a cup beneath the telescope, which fits a dome on the head of the tripod. A screw passing through the dome into the cup is tightened to clamp the spherical surfaces together (Fig 7.2).

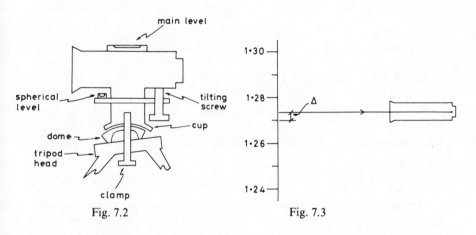

Fig. 7.2   Fig. 7.3

7.9 The parallel-plate micrometer

In general, the line of sight will intersect a levelling staff between two graduations (Fig 7.3). The linear value $\triangle$ must be interpolated between the graduations if readings to this accuracy are required. In Fig 7.3, the staff reading would probably be recorded by estimation as 1.273 m. If a parallel sided glass block is placed in front of the objective of the level telescope, and rotated about a horizontal axis through its centre and perpendicular to the line of sight, the distance $\triangle$ can be measured by making use of the principle described in section 2.4.1.3; if t is the thickness of the glass plate and i its inclination to the vertical,

$$\triangle = t \sin i \, [1 - (1 - \sin\triangle \, i)^{1/2} \, (\mu^2 - \sin^2 i)^{-1/2}]$$

or $\qquad \triangle \simeq ti(1 - \dfrac{1}{\mu})$, i.e. $\triangle \propto$ i approximately

Rotation of the parallel plate about its axis causes the image of the staff seen through the telescope to move up and down. By a suitable rotation, the image of the 1.27 graduation can be made to fall on the intersection of the cross-hairs (Fig 7.4a). Assuming for the moment that $\triangle$ is proportional to i, suppose that the knob used to rotate the plate is graduated in values of $\triangle$ from -0.005 m through zero to $+0.005$ m around an arc of its circumference, where the zero corresponds to the vertical position of the parallel-plate. The reading of this graduated scale against an index then gives $\triangle$ directly.

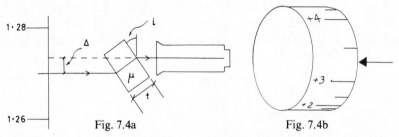

Fig. 7.4a   Fig. 7.4b

In Fig 7.4b, $\quad \triangle = +0.0033$ m and the full reading is
1.27 (staff) + 0.0033 (micrometer) = 1.2733 m

In practice, having the micrometer graduations running from -0.005 m to $+0.005$ m will necessitate in some cases subtracting a micrometer reading from a staff reading. For example, in Fig 7.5a and b the full reading is

1.36 (staff) $-$ 0.0018 (micrometer) = 1.3582 m

Having to add in some cases and subtract in others may introduce an additional source of error. To overcome this, the micrometer scale often runs from zero through $+ 0.005$ (corresponding to the parallel plate in a vertical position) to $+0.010$ m. After obtaining coincidence as before, the micrometer reading is always added to the staff reading.

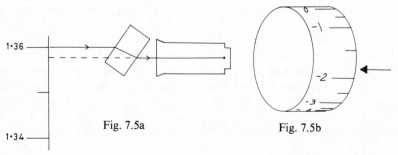

Fig. 7.5a   Fig. 7.5b

Thus in the case illustrated in Fig 7.4a and b the micrometer reading would be

+0.0083 giving a full reading of
1.27 (staff) + 0.0083 (micrometer) = 1.2783 m

and in the case illustrated in Fig 7.5a and b the micrometer reading would be

+0.0032 giving a full reading of
1.36 (staff) + 0.0032 (micrometer) = 1.3632 m

Thus each staff reading is 'in error' by +0.0050 m, but the difference in staff readings is correct, so the derived difference in height is correct.

The foregoing description applies to a level and staff where the staff graduation interval (0.01 m) and the range of the micrometer are the same. The reader should note that it is possible to obtain coincidence between the horizontal hair and only one of the staff graduations. If for example in Fig 7.4a it is attempted to raise the line of sight by tilting the plate backwards, until the 1.28 graduation is on the horizontal hair, it will be found that the mechanism comes against a stop before this situation is reached.

Normal staff graduation intervals are 10 mm and 5 mm, but on some older staffs they can be 0.01 ft and 0.02 ft. The micrometer ranges should be the same. However, it is possible to use a 10 mm micrometer with a 5 mm staff, but not possible to use a 5 mm micrometer with a 10 mm staff. It is important to check that the micrometer and staff graduations are compatible.

On older models, the parallel-plate, micrometer and mechanical linkage are often available as optional accessories. They are fitted to the outside of the telescope and therefore liable to deteriorate in accuracy because of rain, dust etc. and to become damaged accidentally. More modern precise levels are built with the parallel plate and associated optical and mechanical components integral with the telescope and viewed through the same eyepiece. This leads to an improvement in reliability and accuracy.

7.10 Automatic levels

The desirability of producing a telescope for which the line of sight is automatically stabilised in a horizontal plane was recognised over 200 years ago. Early attempts entailed the suspension of the telescope itself so that under the influence of gravity the line of collimation became horizontal. Such attempts failed to produce a system which was both accurate and robust; the mass which had to be suspended was large, damping was difficult and it was not possible to produce suspensions with negligible friction.

The provision of a compensating system inside the telescope itself overcomes some of these problems, but it was not until the beginning of this century that a recorded attempt to produce a stabiliser within the telescope was made (at Carl Zeiss by Heinrich Wild). The first automatic level to go into production was the Zeiss (Oberkochen) Ni 2 in 1950. Other manufacturers followed a few years later and at the present time, almost every manufacturer produces at least one model of an automatic level. Such levels are available for all types of levelling ranging from that required on building sites, to first-order levelling of high precision covering large areas.

Automatic levels have not been readily accepted everywhere, possibly because the surveyor or engineer using one feels that the instrument is out of control; setting a bubble central using a tilting screw induces a feeling of confidence (not always justified) that the 'correct reading' is thereby obtained, but simply setting the line of sight roughly horizontal and then allowing the compensator to take over is a different matter.

On economic grounds, automatic levels have been shown (Förstner, 1953) to be quicker than equivalent tilting levels. The following table shows the speeds of levelling (S) in kolometres per hour using two staffs and one level for different sighting distances (D) in metres.

Type	D (m)	20	40	60	80	100	120
Automatic	S (km/h)	1.2	2	2.4	2.5	2.6	2.7
Tilting	S (km/h)	0.6	1.1	1.5	1.7	1.8	1.9

Similar differences occur for different combinations (e.g. 1 level and 1 staff, 2 levels and 2 staffs etc.). The cost of an automatic level is not significantly higher than that of an equivalent tilting level, and high accuracies (chapter 10) can be attained if suitable precautions are taken, but these precautions are not much more numerous than those necessary if traditional methods are used. Becker, 1977 gives details of the greatly improved economy of using motorised levelling over conventional methods. See also chapter 10.

7.10.1 Basic principles of compensation

The standing axis is set approximately vertical using footscrews or a quickset head and a spherical bubble. There will therefore be a small residual error α (a few minutes in magnitude) in the verticality of the standing axis measured in the direction of the line of sight (Fig 7.6). The compensator must ensure that all rays entering the objective from

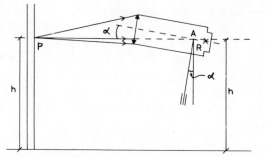

Fig. 7.6

a point P on the staff at the same height as the anallactic point (A) of the telescope are brought to a focus at the intersection R of the cross-hairs. It has been shown (section 2.1.9) that the anallactic point moves laterally along the telescope axis as the internal focusing lens moves. This will cause a variation in the height of the anallactic point if there is an inclination α, and hence there will be a need for the compensation to vary with the sighting distance. For the time being, it is assumed that all sighting distances are infinite and the telescope is at infinity focus. Further consideration of this problem is given in section 10.1.2.

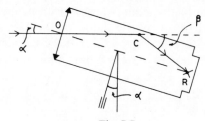

Fig. 7.7

Figure 7.7 shows an incident horizontal ray entering the objective through its optical centre, O. In order to image this ray at R, it has to be deviated by the compensator C through an angle β.

Thus if compensation is to be effective for different residual tilts α, β must be given by

$$\beta = n\,\alpha$$

where n is a constant, often called the enlargement factor.

Taking infinity focus only, if f = OR (the focal length of the objective and focusing lens combined) and if CR = s, then application of the sine rule to the triangle OCR (which has small angles α and β) gives

$$\frac{\beta}{f} \simeq \frac{\alpha}{s}, \text{ so } \beta \simeq \frac{f}{s}\,\alpha \text{ and } n \simeq \frac{f}{s}$$

If the compensator is at the objective, s = f and n has to be unity. If the compensator is mid-way between the objective and the reticule, then s = f/2 and n = 2. Thus as the compensator is placed nearer to the reticule, so the value of n has to increase. The closer the compensator is to the reticule, the smaller it need be, since the cone of rays incident upon it is smaller. Therefore, the weight is less, and the effect of shock will be less. At the same time, damping is easier for a lighter pendulum.

7.10.2 Free suspension compensators

Most compensators consist basically of a suspended prism which acts as a mirror. As a simple illustration of this principle, suppose firstly there are two fixed mirrors M_1 and M_2 in a telescope (Fig 7.8). A horizontal ray entering the objective through its optical centre O continues undeviated until it meets the mirror M_1 from which it is reflected vertically to mirror M_2. From there the ray is reflected horizontally and if the optical path from O to the reticule R is equal to the focal length of the objective (for infinity focus) an image of the object is formed there.

The equivalent optical system is derived by replacing any mirror and the rays reflected from it by the image of those rays in that mirror. Thus the equivalent optical system for the arrangement in Fig 7.8 is derived firstly by replacing M_2 (Fig 7.9a) and secondly by replacing M_1

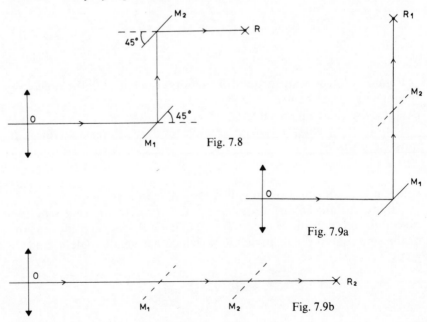

Fig. 7.8

Fig. 7.9a

Fig. 7.9b

(Fig 7.9b) where R_1 is the image of R in M_2 and R_2 is the image of R_1 in M_1. Suppose now that the level illustrated in Fig 7.8 has a residual inclination (α) of the standing axis so that the optical axis is inclined at an angle α (Fig 7.10). An incident horizontal ray through the optical centre will be brought to a focus at R', away from the intersection (R) of the cross-hairs. The equivalent optical system is shown in Fig 7.11, where R_2' is the equivalent image and R_2 is the equivalent intersection of the cross-hairs. R_2 O is the equivalent optical axis and R_2' O is the equivalent horizontal line of sight.

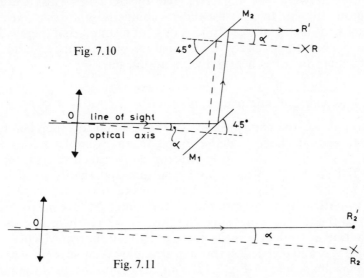

Fig. 7.10

Fig. 7.11

Suppose now that mirror M_1 is freely suspended so that it maintains its 45° inclination to the vertical, as shown in Fig 7.12. The horizontal ray incident upon it will be deviated through an angle 2α from its direction as shown in Fig 7.10. Thus instead of diverging at an angle α, it will converge at an angle α towards the optical axis. If the optical path for this ray from C to R is equal to that from O to C, then this ray will form an image at the intersection (R) of the cross-hairs. The

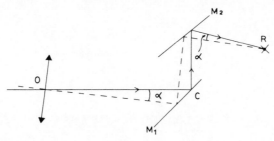

Fig. 7.12

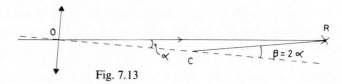

Fig. 7.13

equivalent optical system is illustrated in Fig 7.13. This figure illustrates more clearly that, given a compensator with an enlargement factor 2 (for example a freely-suspended mirror) it must be positioned mid-way between the objective and the reticule. This feature is common to all freely suspended compensators; they give an enlargement factor of exactly 1, 2, or 4 for example and thereby have to be positioned precisely. By the nature of the compensation, the enlargement factor is nearly always an integer.

7.10.3 Mechanical compensators

These rely for the compensation of the residual tilt upon combined optical and mechanical deflections. As an example of this type, suppose that a prism acting as a mirror (M) is attached to the top end of a flat spring (S) which has its bottom end rigidly fixed to the telescope housing. Figure 7.14 shows the situation when the standing axis is vertical and the optical axis horizontal. M' is a mirror rigidly fixed to the telescope housing. Suppose now that there is a residual inclination α. Under the influence of gravity, the mirror M will deform the spring S and will deflect through an angle δ relative to the telescope housing, illustrated in Fig 7.15. This deflection δ causes the image-forming central ray to be deviated through an angle $\beta = 2\delta$, and if compensation is to be accurate, $\beta = n\alpha$, so that $\delta = n\alpha/2$.

In order to determine the required value of the enlargement factor n, consider the forces acting on the compensator unit. In Fig 7.16, G is the centre of mass of the prism unit and P is the point at which the spring is

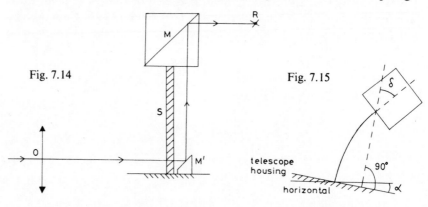

Fig. 7.14

Fig. 7.15

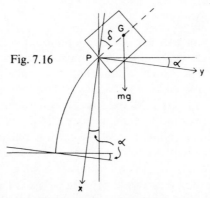

Fig. 7.16

rigidly connected to the prism assembly. Let PG $= s$, and take rectangular x and y axes, inclined at α to the vertical and horizontal respectively, with origin at P.

The forces acting on the prism assembly must be in equilibrium, so if F_x and F_y are the components of the force acting on the end P of the spring, then $F_x = mg \cos \alpha$ and $F_y = mg \sin \alpha$, where mg is the weight of the prism assembly.

Moreover, if M_P is the resultant moment about P, the condition for equilibrium is

$$M_P = mgs \cos \alpha \sin \delta + mgs \sin \alpha \cos \delta$$
$$= mgs \sin (\alpha + \delta)$$

The end P of the spring is therefore subjected to a force with components F_x and F_y and a moment M_P about P. The equation of the curve assumed by the spring can be found in the following way.

For any point (x, y) on the curve, assuming small deflections

$$EI \frac{d^2y}{dx^2} = - F_x y - F_y x - M_p$$

therefore, $\dfrac{d^2y}{dx^2} + \dfrac{mgy \cos \alpha}{EI} = \dfrac{mgx \sin \alpha}{EI} - \dfrac{mgs \sin (\alpha + \delta)}{EI}$

This is the differential equation of the curve which can be solved to give y as a function of the constants m, g, α, E, I, s, δ and L (the length of the spring) and the variable x. Differentiation of this function with respect to x gives the slope and putting $x = y = O$ gives tan δ. Thus it is possible to obtain another relation between δ and α as well as the previous $\delta = n\alpha/2$. In this way n can be determined as a function of m, E, I, L and s and values of these quantities then chosen to give the required enlargement factor, for a given position.

A rather more detailed account is given by Ellenberger, (1957).

7.10.4 Damping systems

A damping system must be efficient, independent of temperature and light in weight. For these reasons, most manufacturers use air damping whereby the compensator is connected to a piston moving in a cylinder. As the compensator oscillates, the piston expels air from the cylinder, the resultant partial vacuum giving rise to an atmospheric force opposing the motion. Other damping systems are magnetic in character. A compensator with a short natural period of oscillation (and short equivalent length) will be more easily disturbed by external forces such as wind and vibration. Thus it is desirable to construct compensators with long equivalent lengths. This is generally done by arranging for the centre of gravity of the compensator to be as far as possible from the point of suspension.

8 Features of modern levels

In this chapter, certain components of modern instruments are described in order to illustrate how the principles of construction outlined in the preceding chapter are put into practice.

8.1 Compensators in automatic levels

There are almost as many different compensators as there are automatic levels. Only a few are described in order to differentiate between the main types.

8.1.1 The Zeiss (Oberkochen) Ni 2

The compensator unit is illustrated in Fig 8.1 and Fig 8.2 shows its disposition inside the telescope. Because of the inclination of the wires, it is known as the V-type compensator.

Compensation is achieved by a combined mechanical and optical enlargement.

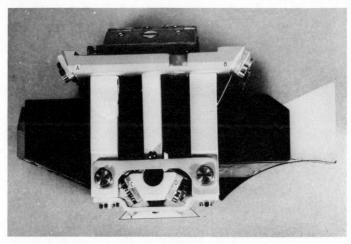

Fig. 8.1

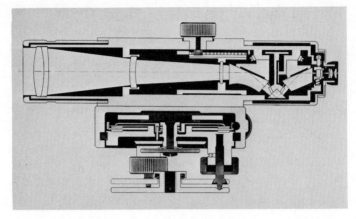

Fig. 8.2

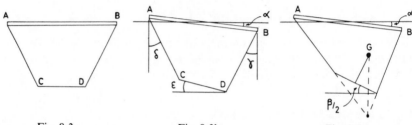

Fig. 8.3a   Fig. 8.3b   Fig. 8.3c

Figure 8.3a shows the suspension frame ABCD; section AB is attached to the telescope barrel and the reflecting prism is attached to section CD. The connection of C to A and of D to B is by wires. If a residual inclination, α, exists (Fig 8.3b) then from geometrical considerations alone,

$$AB \sin \alpha + BD \cos \gamma = AC \cos \delta + CD \sin \epsilon$$

But $AC = BD$ and $\delta \backsimeq \gamma$, so $AB \sin \alpha \backsimeq CD \sin \epsilon$

Because α and ϵ are small, $\sin \alpha \backsimeq \alpha$ etc, so

$$\frac{\epsilon}{\alpha} \backsimeq \frac{AB}{CD}$$

With a deflection ϵ of the prism, the incident ray will be deflected $2\epsilon (= \beta)$. Thus the enlargement factor, n, is given by

$$n = \frac{\beta}{\alpha} = 2 \frac{AB}{CD} = 6 \text{ if } AB = 3 \text{ CD}$$

However, no account has been taken of the weight of the compensator. When this is introduced (Fig 8.3c) by assuming all the weight is at G, the resultant deflection of the prism is greater than ϵ in Fig 8.3b with the result that the enlargement factor is increased from 6 to 7.4.

The weight of the suspended unit is 20 g but each suspension wire has a load-bearing capacity of 2 kg. The equivalent length is 130 mm corresponding to a natural frequency of 1.4 Hz. With damping, the oscillations die down within 0.5s.

8.1.2 Filotecnica Salmoiraghi

A suspended objective is used on the earlier models and a suspended reticule in later models. In each case, the enlargement factor is unity. Figure 8.4 illustrates the basic construction of one of the earlier models.

A horizontal ray enters the periscope-type telescope through a plane glass cover plate (1), is reflected vertically downwards by the prism (2) and passes through the optical centre (O) of the objective (3). This central ray is then reflected horizontally by the prism (4) and falls on the intersection of the cross-hairs (R).

The objective is suspended from S, where SO is equal to its focal length, f. For infinity focus, the optical path from O to R is also equal to f.

Now suppose the standing axis has a residual inclination α (Fig 8.5a). In this figure, the prism (4) is replaced by its optical equivalent and it is assumed to begin with, that the objective is fixed to the case of the telescope, and not allowed to hang freely from S. An incident horizontal ray will thus be brought to a focus at R′, away from the intersection of the cross-hairs.

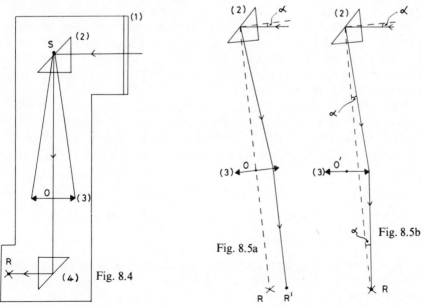

Fig. 8.4

Fig. 8.5a

Fig. 8.5b

Now suppose the objective is allowed to hang freely from S so that its optical centre is at O′ (Fig 8.5b). The incident horizontal ray will make an angle of 2α with the optical axis of the lens and will therefore be refracted as shown and brought to a focus at R, giving the required stabilisation.

8.1.3 Rank (Hilger & Watts) Autoset

Fig. 8.6

The compensator uses two freely suspended reflecting surfaces, each producing a deflection 2α of the incident ray. The enlargement factor is 4 therefore and the compensator must be placed close to the reticule. Figure 8.6 shows the compensator, which has two suspended prisms (1) and (3) and one fixed prism (2).

Figure 8.7 shows the path of an incident horizontal ray entering the optical centre (O) of the objective when there is no residual tilt and the optical axis is horizontal. The image formed by this central ray is at R, the intersection of the cross-hairs after reflection at prisms (1), (2) and (3).

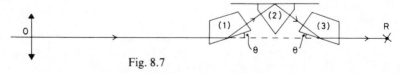

Fig. 8.7

Now suppose (Fig 8.8) there is a residual inclination α and that all three prisms are fixed to the telescope barrel. The divergence of the path is constant (α), with the result that the image is formed at R′.

Fig. 8.8

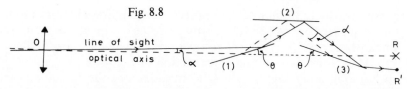

Now suppose that the prisms (1) and (3) are allowed to assume their natural inclination θ to the horizontal. In Fig 8.9, all three prisms are replaced by their optical equivalents.

Fig. 8.9

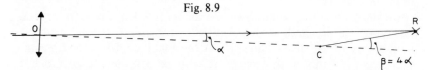

A jewelled pivot or a fine knife-edge pivot would not be satisfactory; friction must be absent and yet robustness is essential. The suspension used therefore is a flexure pivot which by correct design of the dimensions enables the restoring couple after deflection to be zero, thus the suspension behaves as a frictionless pivot. The safety factor is 350 to 1 (i.e. the ratio of breaking force to weight suspended) so robustness is also achieved. The system is air-damped.

8.1.4 The Zeiss (Jena) Ni 007

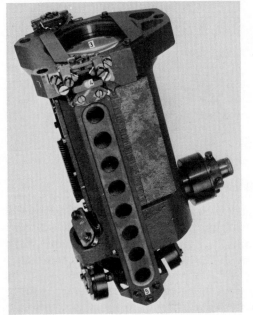

Fig. 8.10

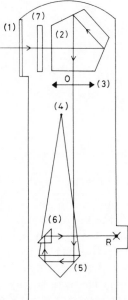

Fig. 8.11

Externally, this is similar to the Filotecnica Salmoiraghi level in so far as it has a periscope appearance. However, the compensator is not the same in each case. The Ni 007 compensator is shown in Fig 8.10 and Fig 8.11 illustrates the basic construction of the level.

A horizontal ray entering the objective (3) along the optical axis via the sealing glass (1) and pentaprism (2) is refracted by the prism (5) suspended from a pivot (4), is then refracted by the fixed prism (6) and is imaged on the intersection of the cross-hairs (R). (7) is a parallel plate micrometer.

Suppose a residual inclination α exists (Fig 8.12). For infinity focus, the image is formed at R', a linear distance a = RR' from the intersection of the cross-hairs. If f is the focal length of the objective, then a $\simeq$ fα. Thus if the image R' can be displaced a distance 'a', compensation is achieved. This displacement is made by the suspended prism (5). If this is displaced normal to the incident ray (Fig 8.13) through a distance a/2, the ray will be displaced a distance 'a'. Thus if compensation is to be complete, the prism (3) must swing through a distance a/2 relative to the telescope when this is inclined at an angle α. This will happen if the prism (5) is suspended a distance f/2 below point (4) which must be at the same height as the optical centre of the objective.

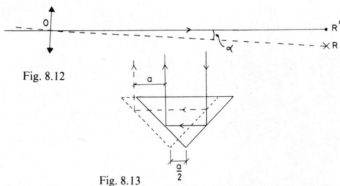

Fig. 8.12

Fig. 8.13

8.1.5 The Ertel INA and BNA 65

The compensator in these levels is of the type described in section 7.10.3 (the upright astatic pendulum). The compensator is located as shown in Fig 8.14. A horizontal ray entering the optical centre (O) of the objective (1) is refracted by prism (2) into prism (3) which is supported on a spring (4) rigidly attached at its base to the body of the level (5). The central ray is then reflected by the rhomboid prism (6) onto the intersection of the cross-hairs (R). If a residual inclination, α, is present, without a compensator the image would be formed at R'

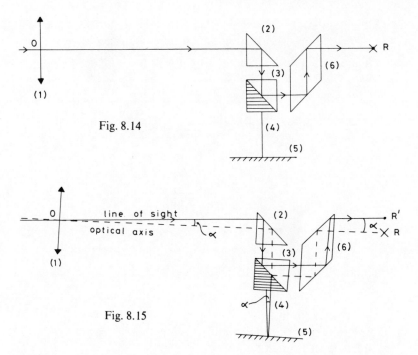

Fig. 8.14

Fig. 8.15

(Fig 8.15). Compensation is achieved by allowing the prism (3) to deflect through an additional angle α (Fig 8.16). Thus the deflection produced is $2\alpha = \beta$ and the enlargement factor $n = \beta/\alpha = 2\gamma/\alpha$. For an enlargement factor $n = 3$, and at the maximum deflection of 20', the safety factor is 50. Magnetic damping is used.

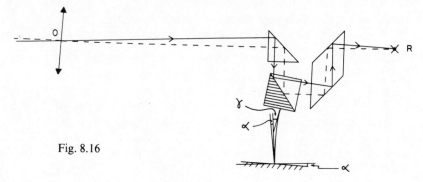

Fig. 8.16

8.2 The use of micrometers in levels

8.2.1. The Wild N3 Precision Level

Figure 8.17 illustrates the older version of the Wild N3.

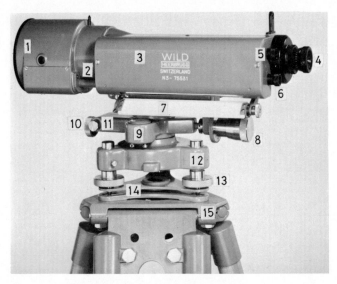

Fig. 8.17

1. parallel plate housing
2. linkage from parallel plate to micrometer drum
3. spirit level housing
4. main telescope eyepiece
5. bubble viewing microscope eyepiece
6. micrometer reading microscope eyepiece
7. reflecting plate for illuminating spirit level
8. tilting screw
9. spherical spirit level
10. clamp for rotation about standing axis
11. slow motion screw for rotation about standing axis
12. tribrach
13. footscrew
14. trivet stage
15. tripod head.

Figure 8.18 shows examples of the views through the three eyepieces. The main spirit level is central, the staff reading is 1.48 m and the micrometer reading is 0.00653 m giving a full reading of 1.48653 m.

The newer version of the Wild N3 Precision Level is illustrated in Fig 8.19 which shows, by comparison with Fig 8.17, that the incorporation of the parallel plate and other mechanisms inside the main body of the instrument gives much less chance of accidental damage and mechanical wear from dust, etc. This is a feature of modern precise levels and has been a major factor in the improvement

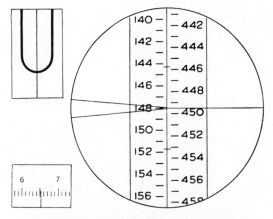

Fig. 8.18

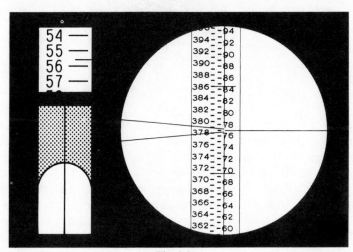

Fig. 8.19

Fig. 8.20

of accuracy and reliability of precise levels in recent years. The fields of view through the eyepieces of the newer N3 are illustrated in Fig 8.20. It shows the bubble coincidence and a reading of 77 cm (staff) and 0.556 cm (micrometer) equivalent to a full reading of 0.775 56 m.

8.2.2 The Zeiss (Jena) Precision Automatic Level Ni 002

This is shown in Fig 8.21. The ray paths are illustrated in Fig 8.22a, from which it can be seen that the optical system is fundamentally different from those of nearly all other precise automatic levels, the exception being the MOM NiA3 system which it resembles both in appearance and operating principles.

Fig. 8.21

In the Zeiss (Jena) Ni 002, rays from the staff enter the instrument through the cover glass (1) and are converged by the objective lens system (2) towards a suspended plane mirror (3) from which they are reflected to a reticule (4) at the objective and in the focal plane of the objective/mirror combination. Images of staffs at different sighting distances from the level are focused at the plane of the reticule by longitudinal displacement of the plane mirror (3). The observer views the image (and the reticule) by means of the telesystem (14) and the eyepiece (13).

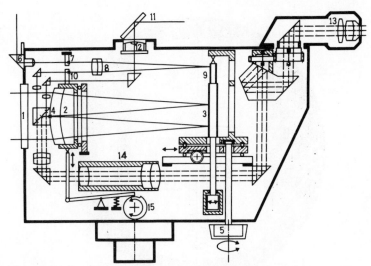

1. cover glass
2. objective
3. suspended plane mirror
   compensator and focusing device
4. reticule
5. compensator reversal knob
6. illumination prism
7. micrometer index
8. micrometer objective lens
9. plane mirror
10. micrometer scale
11. illumination mirror
12. spherical level
13. eyepiece for object,
    micrometer and spherical
    level
14. telesystem
15. micrometer knob

Fig. 8.22a

Instead of having a parallel plate in front of the objective, the Ni 002 has a micrometer scale (10) attached to the objective which can be moved transversely relative to a fixed index (7). Rotation of the micrometer knob (15) causes the objective/reticule/micrometer scale combination to move transversely until the appropriate staff graduation is bisected by the horizontal cross-hair. By means of the illumination prism (6), the micrometer objective (8), the auxiliary plane mirror (9) and the same telesystem (14) as is used for viewing the staff, the micrometer scale and index are presented in the field of view of the eyepiece (13). The observer can also verify at the time of reading the staff that the spherical level is centred. This is done by illuminating the bubble by the plane mirror (11) and by viewing it through the telesystem (14) and eyepiece (13). Figure 8.22b shows the field of view through the eyepiece of the Ni 002. The spherical bubble is on the left and the micrometer scale and index on the right. The staff graduations

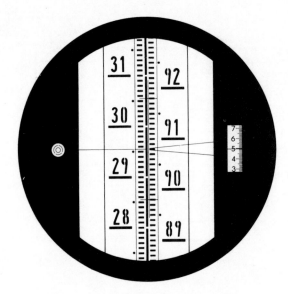

Fig. 8.22b

are at intervals of 10 mm. The reading is 908 cm at the staff and 0.495 cm (to the nearest 0.005 cm) at the micrometer index, giving a full reading of 9.08495 m. There is a false zero on the staff.

The optical system of the Ni 002 has been designed to reduce significantly a number of the errors which occur in automatic levels and which are described in chapter 10. Hüther, 1973 describes the principle of compensation and Scheufele, 1977 gives further details. In addition, its use in motorised levelling is made easier by the swivelling eyepiece which allows the observer to take forward and backward sights without having to move around the level. The focusing, micrometer and slow-motion knobs and the coarse sighting device are situated on both sides of the level so that they can be used easily from the same operating position.

8.3 Quickset head

Preliminary setting-up is made easier and quicker by the use of a 'quickset' head instead of footscrews. Figure 8.23 shows the Hilger & Watts (Rank) quickset head. A dome (1) bolted to the tripod accepts a cup (2) attached to the base of the level. A bolt (3) through the dome connects with the cup. Before this bolt is tightened, the level is rocked on the dome until the spherical level is centred. Then the bolt is tightened so that the cup is rigidly located on the dome.

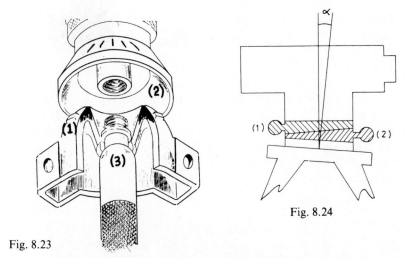

Fig. 8.24

Fig. 8.23

A recent development is the use of contra-rotating wedges by Zeiss (Jena) on the Ni 050. These are illustrated diagramatically in Fig 8.24, as the shaded parts (1) and (2). These are rotated in opposite directions by the small handles until the residual inclination α of the tripod head is annulled.

8.4 The dumpy level

Fig. 8.25

Figure 8.25 shows the Wild NK01 dumpy level with horizontal circle, graduated 0 g to 400 g. The footscrews and bubble allow the standing axis to be set vertically. The line of sight is then horizontal for all pointings which is the characteristic of the dumpy level.

8.5 Rotating laser levels

8.5.1 The Spectra-Physics 944 LaserLevel SL

This is illustrated in Fig 8.26. It can also be mounted on a tripod. The source of the radiation is a HeNe laser with a maximum power output of 5mW. The collimated laser beam is refracted through a pentaprism which rotates about a vertical axis. The beam is 10 mm wide at the instrument. Electrolevels mounted orthogonally detect tilts from the vertical direction and the resultant electrical signals are used to drive correction motors which restore the verticality of the axis of rotation. It is necessary to set the instrument level only to within 8° for the

Fig. 8.26

stabilisation to be effective. It is usually necessary to wait about 1 1/2 minutes from switching on before the stabilisation is complete. In this time, uniform temperatures are achieved in the electrolevels. If the instrument is accidentally moved outside the 8° range of automatic stabilisation, the laser is automatically switched off. A green light shows to tell the operator that the beam is horizontal. The accuracy of stabilisation of the horizontal plane is about ± 10″. A 12V D.C. battery giving 15A/hr will be sufficient for a normal day's work.

The instrument can also be used to define a vertical plane. A trivet is attached to the side of the casing and used in conjunction with a small spherical level to set the rotation axis within 8° of the horizontal. The accuracy of stabilisation of a vertical plane is about ± 16″.

The speed of rotation can be varied by means of a setting knob from zero (when the beam can be used to define a direction – horizontal, vertical or at a pre-set gradient) to 300 revolutions per minute. When the rate of rotation is set to zero, the beam can be set to point in any horizontal direction.

The plane defined by the rotating laser beam can be detected at a staff or other target either by eye or by optoelectronic methods. The manufacturers produce a staff and detector combination where the detector tracks up and down the staff until the laser signal is detected. Random atmospheric effects are averaged and the detector comes to rest at the centre of the received signal. An audible signal is emitted when the detector has locked on to the laser beam. A hand-held detector is also produced. These detectors can be used up to a range of about 300 m from the laser level under normal atmospheric conditions.

Descriptions of some of the errors in a rotating laser level are given in section 10.8.

8.5.2 The Gradomat XL

This is illustrated in Fig 8.27. A 12V D.C. power source is used to generate a HeNe laser of 632.8 nm wavelength with average power output of 2.5 mW. The diameter of the beam is 19 mm at the exit aperture.

Two gradients can be pre-set by the user. The major slope is in the direction of the handle of the instrument (which also acts as a sight for aligning the level) and the cross slope is at right angles to the major slope. Each gradient can be set at intervals of 0.01% from zero up to 5%. The speed of rotation of the laser beam can also be pre-set, at intervals of 100 r.p.m. from zero to 600 r.p.m. These pre-set values of gradients and speed are entered at a keyboard and stored in the memory of a small controlling microprocessor. A luminescent diode

Fig. 8.27

display indicates to the user which values have been pre-set and stored. Initially, the instrument takes about a minute to stabilise. During this period, a warning light is illuminated on the face of the keyboard and the laser is not switched on.

During operation, any accidental departure from the horizontal (or from the pre-set gradients) of the laser results in an output current from the potentiometers associated with the electrolevel vials. These outputs are used to drive correcting motors to restore the correct gradients. For such closed loop servo systems, there is a conflict between stability and accuracy. Generally, if high stability is required, there is a loss of accuracy and the response is sluggish. If high accuracy is required, the response is more sensitive to random influences (or noise) and the stability is lower. Whether stability or accuracy is more important will depend on the use to which the level is put. On later models (the XLC for instance) it is possible to adjust the potentiometers towards higher or lower accuracy.

The range of automatic compensation of tilts is 6% (about 3 1/2°) but if the error of the alignment is more than ± 0.05%, the laser is automatically switched off. It might be necessary for example to use the level close to heavy vibrating machinery, when the random tilts transmitted to the level could frequently move it outside the ± 0.05% tolerance. To avoid the resultant intermittent radiation, it is possible to extend the tolerance for use under those conditions, but the effects of the vibrations will nevertheless introduce random errors in the alignment greater than the normal ± 0.05%. The main reason for automatically switching off the laser is to prevent it being used if the level has been accidentally knocked out of alignment.

The pre-set speed of rotation is stored in the memory of the microprocessor. As the rotator motor turns, its velocity is recorded by an optoelectronic coupler (section 2.4.2.4) consisting of an infra-red LED and a phototransistor. Light from the LED falls on the teeth of one of the gear wheels of the rotator motor and is reflected to the phototransistor which produces electrical pulses. The frequency of these pulses represents the speed of rotation of the laser. The microprocessor is used to derive and compare this measured speed with the pre-set value stored in its memory. Any discrepancy results in an increase or decrease of the current to the rotator motor until the measured speed is correct. When zero speed is pre-set, the beam can be rotated either clockwise or anticlockwise by depressing one of two keys on the keyboard. By these means, the laser can be used to define a direction. The instrument can be used also in a vertical mode to define a vertical plane or direction.

A laser detector (called the Laser Scout) is produced by the manufacturers of the Gradomat XL. It can be hand-held, or attached to a staff or to an excavator. It consists essentially of detector circuits and a display circuit. A detector circuit receives the photocurrent generated by the incident laser and compares the voltage with a reference voltage to test whether the received signal is significant. If the generated voltage is greater than the reference voltage, a current is sent to the display circuit. There is one detector circuit from each of two photodetectors, one higher than the other. If both circuits send a signal to the display circuit, it indicates to the user that the laser beam is in the centre of the array of detectors. If the upper detector circuit only sends a current to the display circuit, it indicates to the user that the laser beam falls just above the centre of the detectors, and conversely for the the lower detector circuit. If neither sends a current to the display circuit, it indicates to the user that the laser beam is not near the unit, and the unit must be moved vertically upwards or downwards in order to detect the laser. The indications to the user are both visual, by coloured lamps, and aural, by intermittent or steady tones.

The user can increase or decrease the sensitivity of the detector by setting a lower or higher reference voltage respectively for the detector circuits. Increasing the voltage sensitivity means that the accuracy of location of the centre of the laser beam is decreased – it is possible to move the detector vertically across the beam and still obtain a signal from each detector circuit. The sensitivity can be adjusted to allow the centre of the beam to be detected with accuracies between 2 mm and 100 mm at 30 m from the level.

9 Adjustments and accuracies of non-automatic levels

As with theodolite adjustments, the adjustments to levels should always be made in accordance with the manufacturer's instructions. However, certain basic principles apply to all non-automatic levels and the following descriptions are in general terms. Adjustments to the telescope (eyepiece-setting and elimination of parallax) are the same as those already described in sections 4.5.1 and 4.5.2.

9.1 Adjustment of the tilting level (collimation)

The requirement is for the line of sight to be parallel to the principal tangent of the spirit level. Then, when the bubble is central, the line of sight is horizontal. A procedure for testing and adjusting simple levels to be used for low-order work follows. For other levels, the test and adjustment should be as described in section 9.5.1.

1. Select two points A and B, about 75 m apart and on fairly level ground. Put in firm ground marks at A and B.
2. Set up the level midway between A and B (pacing is sufficiently accurate to determine the mid-point). Direct the telescope towards a staff held vertically on A, centre the bubble using the tilting screw, eliminate parallax and take a staff reading (R_A).
3. From the same position, take a similar reading to the staff held vertically on B (R_B).
4. Remove the level and set it up about 1 m from the staff at A. Take a staff reading on A (R'_A).
5. From the same position, take a reading on the staff at B in the normal way (R'_B).
6. Calculate the difference in height from A to B from the two sets of readings;

$$\triangle h = R_A - R_B$$

$$\triangle h' = R'_A - R'_B$$

If $\triangle h = \triangle h'$, the adjustment is satisfactory. $\triangle h$ is the true difference in height between A and B, even if the level is not in adjustment and it is found that $\triangle h \neq \triangle h'$. In Fig 9.1, the line of sight is assumed to be elevated through an angle α above the horizontal, so that each reading is too high by an amount x, hence $R_A - R_B$ is the correct difference in height. When observations are made from the second position of the level (Fig 9.2) the difference in height ($\triangle h'$) will not equal the true difference, $\triangle h$, because the reading R'_A is not in appreciable error whereas the reading R'_B is in error by $+2x$.

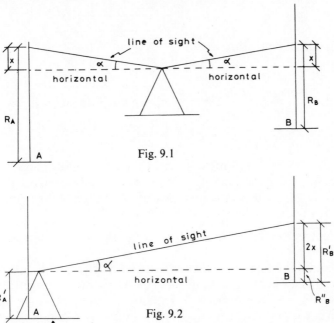

Fig. 9.1

Fig. 9.2

7. If $\triangle h \neq \triangle h'$ then calculate $R''_B = R'_A - \triangle h$. This is the correct reading for the staff at B (i.e. $R'_B - 2x$).
8. By means of the tilting screw, bring the horizontal hair on to the staff at this reading.
9. The bubble will no longer be central, so re-centre it using the bubble locking screws.
10. Repeat the test until $\triangle h = \triangle h'$ within 5 mm. The adjustment is then satisfactory for ordinary levelling.

This adjustment is often referred to as the *two-peg test*. A numerical example follows:

1. From the first position,

$$R_A = 1.872 \text{ m}$$
$$R_B = 1.264 \text{ m}$$
$$\text{(True) } \overline{\triangle h = + 0.608 \text{ m}}$$

2. From the second position, R'_A =   1.380 m
                                R'_B =   0.698 m
                           $\Delta h'$ = + 0.682 m

$R''_B = (1.380 - 0.608) = 0.772$ m (adjusted)
(Note that the line of collimation was depressed by 0.074 m over the distance AB).

3. From the 1st position again R_A =   1.926 m
                                 R_B =   1.318 m
                       (True) Δh = + 0.608 m

4. From 2nd position again R'_A =   1.378 m
                                 R'_B =   0.768 m
                           $\Delta h'$ = + 0.610 m

Thus $\Delta h = \Delta h' - 2$ mm. Adjustment satisfactory.

9.1.1 Adjustment of the spherical spirit level

This spirit level is used in conjunction with either a quickset head or footscrews to set the line of sight approximately horizontal, so that final levelling can be carried out within the range of the tilting screw. The adjustment of the spherical level is as follows:

1. Centre the circular bubble using the quickset head or the footscrews.
2. Rotate the telescope 180° about the standing axis. If the bubble moves off centre, bring it back half-way towards the centre by the bubble locking screws. (Generally, there are three of these, situated underneath the bubble housing. Loosen slightly that screw towards which the bubble is to run, and then tighten the other two).
3. Bring the bubble back the rest of the way to the centre by means of the quickset head or the footscrews.
4. Rotate the telescope through a further 180°. If the bubble remains central, the adjustment is complete. If not, repeat the procedure until the bubble stays central.

Care should be taken not to tighten the adjusting screws too much; this will cause deformation of the vial and eventual fracture.

9.2 Adjustments of the dumpy level

Two adjustments are necessary. First of all, the principal tangent of the spirit level must be perpendicular to the standing (vertical) axis. Secondly, the line of sight must be parallel to the principal tangent of the level vial. Then, the standing axis can be set vertical and the line of sight will describe a horizontal plane when it is rotated about the vertical standing axis.

9.2.1 Adjustment of the spirit level

This is carried out in exactly the same way as the adjustment of a theodolite plate level (section 4.1) using the footscrews and the spirit level locking nuts.

9.2.2 Adjustment of the line of collimation

The procedure described in section 9.1 is followed (except that no tilting screw is used as described in step 2 for example) up to step 7, after which the following procedure is followed.

8. The correct reading (R''_B) is set on the staff at B by means of the reticule locking screws (see section 4.2, step 7) the reticule being moved vertically until the horizontal hair gives the correct reading.
9. Repeat the test until $\triangle h = \triangle h'$ within 5 mm.

9.3 Adjustment of the reversible level

Because the telescope can be rotated 180° about its longitudinal axis, the test for collimation error is more easily carried out than for the simple tilting level. A staff about 75 m away is sighted on bubble-left and a reading taken (R_L). From the same position, the telescope is rotated and a reading (R_R) taken on bubble-right. If the level is in adjustment, $R_L = R_R$. If not, the mean of R_L and R_R is calculated and this reading is set on the level by means of the tilting screw, followed by adjustment of the spirit level as described in section 9.1 (step 9). The test is repeated until R_L and R_R agree to within the desired limits.

9.4 Instrumental accuracies

For a tilting level assumed to be in adjustment, the accuracy of a staff reading depends basically upon the accuracy with which the bubble can be centred (neglecting non-instrumental factors such as atmospheric conditions, length of sight, nature of ground etc.). If a simple mirror is used to view the bubble, setting can be made to about ± 0.4 mm with 2 mm graduations. For a builder's level the value of one bubble division is about 40″ per 2 mm. Thus the line of sight can be levelled to about ± 8″. For a length of sight of 100 m, an 8″ inclination gives an error in the staff reading of about 4 mm.

One step higher in the precision of levels is an 'engineer's level' with the same spirit level, but using coincidence reading. This gives an approximate 10-times increase in accuracy enabling the line of sight to be set horizontal to about ± 1″.

If a level is constructed with a more sensitive spirit level (20″ per 2 mm for example) and coincidence reading is used, horizontality to about ±0.5″ can be achieved and the use of a parallel plate micrometer is justified.

There is no universally recognised classification of levels, but the terms 'builder's', 'engineer's' and 'engineer's precise' levels are generally taken to be those levels giving theoretical setting accuracies of the line of sight of about ± 8″, ± 1″ and ± 0.5″ respectively.

It must not be thought that the mere provision of a more sensitive level vial and bubble viewing device is all that is required to improve the accuracy of a level. Objective aperture, magnification, resolution, image brightness, machining tolerances and insensitivity to temperature changes must also be improved if the full benefits of the increased accuracy in levelling the line of sight are to be realised.

In general, every level (and theodolite) is an assembly of components, each made with only the precision necessary to give the final design accuracy. It is therefore desirable for a prospective user to recognise this fact, to assess the accuracy required in any particular project and to select an instrument which will give that desired accuracy and no more.

The manufacturers often quote the accuracy attainable using normal field techniques and taking reasonable precautions for different levels in the form 'x mm per km' where x can be the average, probable or standard error of a forward and backward run of series levelling over 1 km. Thus, for example, Kern quote ± 7 mm ± 2.5 mm and ± 2.0 mm as mean errors for their GK 0, GK 1 and GK 23 respectively in basic form. These values provide a guide to the accuracy attainable and are of assistance in selecting a suitable level for a particular task.

Kissam (1963) describes a test for estimating the accuracy of a level in which field conditions for series levelling are simulated as closely as possible except that the two staffs remain fixed about 100 m apart and the level is set up at different positions between them. Readings to the two staffs from these different positions enable an estimate to be made of the error in horizontality of the line of sight, together with an estimate of the standard error of the average slope of the line of sight.

This enables a statistical estimate to be made of the standard error per kilometre of levelling, for given lengths of sight. Levels are rated first-order, second-order or third-order according to whether the line of sight can be levelled with a standard error of less than 1.46″, 3.5″ or 6.11″ respectively.

9.5 Geodetic levels

These levels are designed to give a very high accuracy and are used for example to provide the fundamental levelling network on a national basis, to give a framework for large scale civil engineering works, to measure deflections in structures and for the positioning of machinery within very small tolerances. Invariably, they should be used with parallel-plate micrometers and invar staffs if their full accuracy is to be obtained.

Most geodetic levels are not reversible; the introduction of a longitudinal axis of rotation means an additional source of instrumental error, increased weight, instability and cost. With suitable field precautions, the mean error of a double levelling over 1 km is less than 1 mm.

Two adjustments are necessary for a non-automatic geodetic level; the line of sight must be horizontal when the spirit bubble is central, and secondly, the vertical planes containing the line of sight and the principal tangent of the level vial must be parallel. The latter requirement theoretically is one which all tilting levels must meet, but in practice, the existence of such an error in lower-order levels is negligible. Such an error is often called a *cross error*, or *intersection error*.

9.5.1 Adjustment of the spirit level

The procedure is basically that described in section 9.1, but rather more care is needed. In all cases, the manufacturer's instructions should be referred to, but the following procedure illustrates the basic principles of the adjustment.

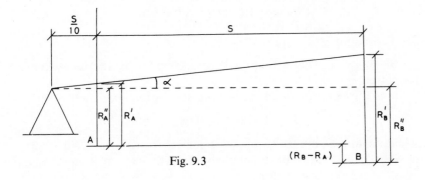

Fig. 9.3

1. Select two points A and B, about 60 m apart. These points should be stable and well defined.
2. Set up the level midway between A and B. The mid-point should be found to an accuracy of about 0.1 m, so taping is preferable to pacing. From this mid-point, take full staff and micrometer readings to A (R_A) and B (R_B). Then ($R_B - R_A$) is the correct difference in level between B and A.
3. Set up the level on the line BA produced at a distance from A equal to 1/10th of the distance AB (Fig 9.3). Take staff readings to A' (R_A) and to B (R'_B). Calculate the reading (R''_B) which would be obtained on B if the line of sight were horizontal;

$$(R_B - R_A) = (R''_B - R''_A)$$

Therefore, $$R''_B = R_B - R_A + R''_A$$

But $$R''_A = R'_A - \frac{1}{10}[(R'_B - R'_A) - (R_B - R_A)]$$

Therefore, $$R''_B = (R_B - R_A) + R'_A - \frac{[(R'_B - R'_A) - (R_B - R_A)]}{10}$$

This should be equal to the reading actually obtained (R'_B).
4. If $R''_B \neq R'_B$, then obtain the correct reading (R''_B) by setting firstly the micrometer and secondly the staff reading using the tilting screw. This will displace the bubble which can be re-centred using the level vial clamping screws.

The reader should note that no account is taken of curvature and refraction in the adjustment, and that there is therefore a residual error in the line of sight. The effect of curvature and refraction on a staff reading is to make it too high by an amount e given by

$$e = \frac{S^2(1 - 2k)}{2R}$$

where   S is the distance sighted,
         k is the coefficient of refraction
         and R is the radius of the earth.
See for example, Cooper, 1974.

Taking $k = 0.07$ and $R = 6370$ km, $e \simeq 6.8S^2 \times 10^{-5}$ mm where S is in metres.

Thus the reading which should be set on the staff at B is

$$R''_B + 6.8 \times 10^{-5} (S^2_B - S^2_A)$$

where S_A and S_B are the distances in metres from the level (in its second position) to A and B respectively. If $AB \simeq 60$ m, then R''_B is to be corrected by approximately +0.33 mm. This is negligible, bearing in mind the fact that in first-order levelling is necessary to balance backsight and foresight distances to ± 1 m.

In the foregoing modification of the two-peg test, it is necessary to assume a value for the coefficient of atmospheric refraction. The standard value is k = +0.07 and this has been used in the numerical example above. However, Angus-Leppan, 1968 has shown that the coefficient of refraction a metre or so above ground level often differs significantly from the standard value and is sometimes negative. To reduce the effect of atmospheric refraction on the adjustment of geodetic levels, the *three-peg test* described by Angus-Leppan, 1971 can be used. This test not only enables the collimation error of the level to be found, but also the coefficient of atmospheric refraction at the time of the test, so that its effects can be removed from the adjustment of the level.

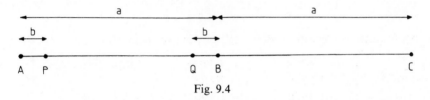

Fig. 9.4

In Fig 9.4, A, B and C are the plan positions of the three pegs and P and Q are successive positions of the level. These five positions are set out beforehand on uniform level ground, with a ≃ 100 m and b ≃ 5 m. The observing procedure is as follows.

1. Set up at P and take readings p_A and p_B to A and B respectively, carefully centring the bubble each time in the case of a tilting level.
2. Set up at Q and take the readings q_A and q_B to A and B respectively, again with care.
3. Compute the correct staff reading from Q to A and adjust the level to give this reading. Compute the correct staff reading from Q to C and check that this is obtained to the required accuracy. If not, repeat the test and adjustment.

To compute the correct staff readings, suppose the collimation error of the level is e radians, the coefficient of atmospheric refraction is k and the radius of the earth is R. Then any staff reading made over a distance S is corrected for curvature, refraction and collimation by subtracting $[Se + S^2(1 - 2k)/2R]$. If H_A, H_B and H_C are the reduced levels of the pegs A, B and C respectively, then, for the observations from P:

$H_B - H_A = [p_A - be - b^2K] - [p_B - (a - b) \ e - (a - b)^2K]$ and
$H_C - H_A = [p_A - be - b^2K] - [p_C - (2a - b) \ e - (2a - b)^2K]$

where $K = (1 - 2k)/2R$.
Similarly, for the observations from Q:

$H_B - H_A = [q_A - (a - b) \ e - (a - b)^2K] - [q_B - be - b^2K]$ and
$H_C - H_A = [q_A - (a - b) \ e - (a - b)^2K] - [q_C - (a + b) \ e -$
$$(a + b)^2K]$$

Thus there are four equations in four unknowns: $(H_B - H_A)$, $(H_C - H_A)$, e and K. After some algebraic manipulation and elimination of $(H_B - H_A)$ and $(H_C - H_A)$, e and K can be found as functions of the staff readings and distances a and b:

$$e = \frac{(p_A - 2p_B + p_C) - (q_A - 2q_B + q_C)}{2(2b - a)}$$

$$K = \frac{(p_B - p_C) - (q_B - q_C)}{2a(2b - a)}$$

An example is quoted where all values are in metres:

$p_A = 1.6208$	$q_A = 1.7081$	$a = 100$
$p_B = 1.5106$	$q_B = 1.6037$	$b = 5$
$p_C = 1.4930$	$q_C = 1.5901$	$R = 6.38 \times 10^6$

Thus, e is found to be -1.0×10^{-5} radians, indicating that the line of collimation was inclined by 2.1" below the horizontal. Also, K is $-2.2 \times 10^{-7} \ m^{-1}$, corresponding to the value of $+1.9$ for the coefficient of refraction, k. If each staff reading is now corrected both for the collimation error and for the effects of curvature and refraction by subtracting $(Se + S^2K)$, the values in the following table are obtained.

Observed staff reading	Dist. S	Collimation error Se	C. & r. effect S^2K	Corrected staff reading
$p_A = 1.6208$	5	−0.0001	negligible	$p'_A = 1.6209$
$p_B = 1.5106$	95	−0.0010	−0.0020	$p'_B = 1.5136$
$p_C = 1.4930$	195	−0.0020	−0.0084	$p'_C = 1.5034$
$q_A = 1.7081$	95	−0.0010	−0.0020	$q'_A = 1.7111$
$q_B = 1.6037$	5	−0.0001	Negligible	$q'_B = 1.6038$
$q_C = 1.5901$	105	−0.0011	−0.0024	$q'_C = 1.5936$

As a check, $H_B - H_A = p'_A - p'_B = +0.1073 = q'_A - q'_B$
and $H_C - H_A = p'_A - p'_C = +0.1175 = q'_A - q'_C$

The method relies on the assumption that the coefficient of atmospheric refraction is constant along the line ABC and remains so during the test and adjustment, hence the need to select a level site on uniform ground and to work carefully but quickly.

9.5.2 Adjustment of the cross-error

The previous adjustment ensures that the line of collimation and the principal tangent of the spirit level lie in the same horizontal plane. There may be, however, an inclination between the vertical plane through the line of collimation and the vertical plane through the principal tangent of the spirit level (i.e. a cross-error). On the assumption that the standing axis is perpendicular to the principal tangent of the level vial, this inclination will not produce an error in the horizontality of the line of collimation; when the bubble is central, the standing axis is vertical and the line of collimation is horizontal. This is so for a dumpy level in perfect adjustment. For a tilting level however, there is no necessity for the principal tangent of the spirit level to be perpendicular to the standing axis. Thus, at each set-up the standing axis is likely to take up a different attitude with respect to the vertical. When the adjustment described in the previous section is carried out, the standing axis has a certain inclination to the vertical. If subsequently the level is set up and the standing axis takes up the same attitude, the line of collimation will be horizontal, even if a cross-error is present. If, however, the standing axis is at an angle α to its direction at the time of adjustment (measured in a direction normal to the line of sight) and the cross-error is δ, then the line of collimation will be inclined at an angle e given by

$$\tan e = \tan \delta \sin \alpha$$

All quantities are small (of the order of a few minutes) so

$$e'' \simeq \alpha'' \, \delta''/\rho''$$

Thus e is of the order of 0.5″ at the worst. It is therefore of significance only in precise levels. The test and adjustment are as follows.

1. Set up about 50 m from a staff with the level orientated so that any two footscrews (A and B) lie on a line roughly perpendicular to the line of sight. Centre the circular bubble using the footscrews.
2. Centre the main bubble using the tilting screw in the usual way and take a staff-reading (R).
3. Turn each of the footscrews A and B by about one revolution but in opposite directions. This inclines the standing axis normal to the line of sight.
4. Using the tilting screw, re-set the reading R.
5. If the main bubble is only slightly displaced, the adjustment can be considered satisfactory. If the displacement is more than the equivalent of about 20″, the bubble should be centred using the horizontally opposed locking screws.

6. Repeat the test and adjustment until the bubble displacement is satisfactory.
7. Repeat the three-peg test described in section 9.5.1 and make any necessary adjustments.

This adjustment is generally unnecessary and it is unlikely that it can be made in the field without difficulty. Nonetheless, if the highest accuracies are aimed at, it is necessary to test for this error and correct it if it is significant.

9.6 Reduction of errors by field techniques

The field techniques employed to reduce instrumental errors will depend upon the accuracy aimed at; the higher the accuracy, the more precautions there are. The following description assumes that a high accuracy is required, and some of these techniques can be ignored for lower-order work.

9.6.1 The effects of temperature on the instrument

The main effect of temperature is to cause a change in the surface tension of the liquid in the vial. If there is a uniform change throughout the length of the bubble, no displacement of the bubble will result. However, if one end of the vial is at a higher temperature than the other, the surface tension decreases at the hotter end with the result that the bubble is displaced towards that end. Thus it is necessary to shade the level from direct sunlight to avoid differential heating. The magnitude of the error is of the order of 1″ for a 0.2°C temperature difference between the ends of a 100 mm bubble vial according to Drodofsky (1956).

9.6.2 The effects of collimation error

It can be seen from section 9.1 that if there is a collimation error of α (i.e. the line of sight is inclined at an angle α to the horizontal when the bubble is central) then the difference between two staff readings is the correct difference in height only if the level is equidistant from the staffs. For this reason it is necessary in high-order levelling for the staff distances to be equalised to within ± 1 m. Such a procedure also eliminates the effects of curvature, and reduces considerably those of refraction. A check on the cumulative totals of back-sight and fore-sight distances should be made to ensure that the difference does not exceed 3m.

In lower-order levelling (to determine spot heights for contouring, or levels for a road scheme for example) it is not economically worth while to balance all the staff distances, especially as several intermediate sights are usually made at each set-up. As precautions against collimation error for this type of work, the level should be tested regularly and adjusted when necessary, staff distances should be balanced for backsights and foresights, and staff distances for intermediate sights should be kept as short as possible and never more than 100 m.

9.6.3 The effects of a change in the instrument height

Suppose the level sinks a distance e_1 after taking a backsight to A and before taking a foresight to B. Then the reading on B will be too low by an amount e_1. At the next set-up, suppose the foresight to C is taken first and then the backsight to B. If the level sinks a distance e_2 between these two readings the reading to B will be too low by e_2. The difference in height between A and C will then be in error by $-(e_1 - e_2)$, that is by an amount equal to the difference between the errors. If however, the backsight is observed first at each set-up, the difference in height between A and C will be in error by $-(e_1 + e_2)$, that is by the sum of the errors. The former procedure is therefore desirable; observations are made to the backsight first at the nth set-up and to the foresight first at the (n + 1)th set-up. With two staffs, this means that the same staff is read first at each set-up, the staffs 'leap-frogging' each other along the line. Also, there should be an even number of set-ups between bench marks.

Tests made on different soils with a Wild N3 level by Belshaw (1959) indicate that the level rises as a result of the slow release of stresses which are set up in the soil by digging in the tripod feet. Figure 9.5 shows the rise plotted against elapsed time for a dry gravel on clay. Dry soils showed slightly more rebound properties than wet.

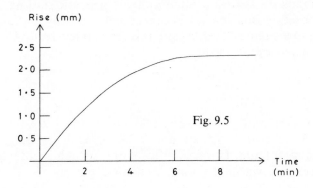

Fig. 9.5

9.6.4 The effects of staff errors

Staffs are often neglected when sources or error are being considered, but several errors can occur.

9.6.4.1 Graduation error

For precise levelling, staff graduation errors can be determined in a laboratory by comparison with a standard tape using a travelling microscope. Corrections can then be applied to observed staff readings. These corrections are generally of the order of 0.1 mm.

9.6.4.2 Zero error

This is present if the zero graduation does not coincide with the base of the staff. If only one staff is used, the error is eliminated at each set-up. If two staffs are used, then the error is eliminated over a pair of set-ups if one staff is used for a backsight for one set-up and a foresight for the other. There should be an even number of set-ups between bench marks.

9.6.4.3 Non verticality

If a staff is not held vertically, then the reading obtained is always too high by an amount $e \simeq 1/2\,S\,\theta^2$ where S is the staff reading obtained and θ is the angle in radians between the face of the staff and the vertical. This is an appreciable error even in low-order levelling (for a staff reading of 2 m and a 3° inclination the error is about 2.5 mm). In lower-order levelling, the staff can be swung backwards and forwards through the vertical and the lowest observed reading taken. Alternatively, a circular spirit level can be used but this must be checked regularly against a plumb-line. A geodetic staff has handles and sometimes struts so that it can be held steadily and vertically, according to a circular level. Again, this circular level must be checked regularly against a plumb-bob.

9.6.4.4 Staff warping

If the face of the staff is not a plane, then the reading is always too high. If the deviation from a plane is about 20 mm at the reading, then the error is about 0.1 mm. This is significant only in higher-order levelling, when checks for warping should be made regularly.

9.6.4.5 Temperature effects

Lower-order levelling with wooden or light alloy staffs is not affected significantly by the thermal expansion of the staff. In higher-order levelling, invar is used which has a coefficient of thermal expansion of the order of $1.3 \times 10^{-6}/^\circ C$. Thus differential expansion errors are insignificant in comparison with other errors such as reading and atmospheric errors. If necessary, they can be allowed for by calculating corrections.

9.6.4.6 Staff sinking

Whether one staff or a pair of staffs is being used, there will be a possibility of one sinking during the time interval between the foresight to it and the backsight to it from the new set-up. Suppose a backsight to A and a foresight to B are made from one level position and then the staff at A is 'leap-frogged' to C beyond B, whilst the level is set-up between B and C. If the staff at B sinks by an amount 'e' between the foresight to it and the backsight to it, the difference in level between A and C will be in error by $-$ e. This error can be reduced by observing the backsight to B as soon as possible after the foresight to it. This however is the opposite of the requirement for reduction of the effects of a change in instrument height (section 9.6.3). Accordingly, the staff positions in higher-order levelling are selected carefully to avoid settling and if all the ground is soft, a base-plate can be used to reduce the pressure of the staff on the ground.

9.6.5 The effects of curvature and refraction

In section 9.5.1 it is shown that the combined effect of curvature and normal refraction is to make a staff reading too high by an amount $e \simeq 6.7 S^2 \times 10^{-5}$ mm where S is the sighting distance in metres. Thus, if at one set-up, the backsight and foresight distances are equal, the difference in the staff readings is correct, provided the coefficient of refraction is the same over both lines of sight. This is likely if the thermal properties of the ground over both distances are the same and if the ground is fairly flat. A danger arises in higher-order levelling where a level line is run up or down a steady slope, when a systematic refraction error arises. In such a case (going up, for example) the foresight will be nearer the ground than the backsight and will pass through warmer air, being refracted less than the backsight ray. Thus the forestaff reading will always be higher than it would be with equal refraction with the result that the reduced level of a point at the top of the slope is lower than it should be. Therefore it is not advisable to level directly up a long slope, but to zig-zag across the line of greatest slope in an attempt to reduce the systematic error.

The foregoing systematic error is present if the temperature is assumed constant with time. If the refraction varies uniformly with time (as it is likely to if the temperature gradually rises or falls) the effects of a systematic error can be reduced if the same staff is always read first.

The effects of curvature are eliminated if sighting distances are equal.

9.6.6 Reading and booking errors

As with other precautions, the steps taken to minimise these errors increase in number with an increase in desired accuracy. For low-order levelling, the booking procedure and arithmetical checks are well known and they are not discussed here. For higher-order levelling, the method of booking and checking varies with the type of staff used, the accuracy aimed at and preferences of individuals. A suggested method for a staff having one set of graduations is as follows. Fig 9.6 shows the order in which the readings are taken at two successive set-ups and

Backsight		Foresight		Distance		Diff. level		Remarks
Stadia	Level	Level	Stadia	Back	Fore	Rise	Fall	
$^+$1	3	4	5			3-4		BM -
2	8	7	6			8-7		CP 1
1-2	3-8	4-7	5-6	1-2 = a	5-6 = b	MEAN		
13	12	11	9 $^+$	13-14 = c	9-10 = d		12-11	CP 1 -
14	15	16	10				15-16	CP 2
13-14	12-15	11-16	9-10	a + c	b + d		MEAN	

Fig. 9.6

Backsight		Foresight		Distance		Diff. level		Remarks
Stadia	Level	Level	Stadia	Back	Fore	Rise	Fall	
1.910$^+$	1.78490	1.49850	1.625	0.250	0.255	0.28640		BM-CP1
1.660	1.78482	1.49826	1.370			0.28656		
0.250	0.00008	0.00024	0.255			0.28648		
1.540	1.43092	1.92804	2.045$^+$	0.220	0.230		0.49712	CP1-CP2
1.320	1.43042	1.92786	1.815				0.49744	
0.220	0.00050	0.00018	0.230	0.470	0.485		0.49728	

Fig. 9.7

how they are subtracted or added whilst Fig 9.7 shows a numerical example.

Such a system provides the following checks:

1. A check on the level hair readings; $(3\text{-}8) = (4\text{-}7) \pm 0.001$ m
2. A check on the difference in level; $(3\text{-}4) = (8\text{-}7) \pm 0.001$ m
3. A check on the staff distances; $(1\text{-}2) = (5\text{-}6) \pm 0.01$
4. A check on the running totals of backsight and foresight distances; $(a + c) = (b + d) \pm 0.02$
5. A check against gross errors; $(3\text{-}4) \approx (1\text{-}5)$.

If the staff has double graduations, readings 7 and 8 are made to the second set of graduations, thus also checking against gross errors.

The bubble should be checked before and after each level hair reading and should be thrown off centre and readjusted between readings 6 and 7, to ensure that readings 4 and 7 are independent.

The marks '+' against readings 1 and 9 are reminders that the same staff is read first at each set-up.

10   Errors and accuracies of automatic levels

Nearly all the instrumental errors characteristic of automatic levels are insignificant for low-order work. By using suitable observing techniques compatible with those outlined in section 9.6, the effects of these errors can be considerably reduced and very high precision can be achieved with instruments having suitable objective aperture, magnification etc. to match the compensator. As in the case of non-automatic levels, an automatic level is designed to give a certain accuracy, and first-order automatic levels have given accuracies of less than 1 mm (mean error) per km of double-levelling. For example Wolter, 1963 quotes a mean square error of $\pm 0.34 \sqrt{L}$ mm where L is the distance levelled in kilometres, for levelling carried out from 1959–1962 over a distance of 1326 km, levelled both ways, using a Zeiss (Oberkochen) Ni2. A more recent example of the accuracies that can be obtained from automatic levels with, at the same time, greatly increased speed of levelling is described by Becker, 1977. In Sweden, Zeiss (Jena) Ni 002 levels were used mounted on tripods, but carried on motor vehicles and lowered to the ground for sighting to staffs which were also carried on motor vehicles and lowered to the ground when readings were required. Between 1974 and 1976, 2 900 km of secondary and 200 km of primary levelling were completed. Secondary levelling (where the allowable maximum discrepancy between forward and reverse levelling was $4\sqrt{L_{km}}$ mm) was carried out over an average of about 12 km per day single levelling, an increase of 100% over the non-motorised method. Primary levelling (where the allowable maximum discrepancy between forward and reverse levelling was $2\sqrt{L_{km}}$ mm) was carried out over an average of about 8 km per day single levelling.

Data were obtained at the level, entered into a minicomputer and automatically checked for discrepancies between level hair and stadia readings before transformation and storage as height differences. Motorised levelling was found to give 'a significant increase in production together with decreased costs and an accuracy equal to and in many cases better than that achieved using conventional methods'. The tolerances quoted above were relatively easy to achieve. It was found that 91% of the secondary work resulted in discrepancies between forward and reverse levelling meeting the requirement for primary levelling and 99% of the primary levelling gave discrepancies less than $1 \sqrt{L_{km}}$ mm.

A number of selected references related to motorised levelling are given by Strosche, 1977. Some theoretical discussions and practical results of motorised levelling are given by Peschel, 1977, Seltmann, 1977, Degenhardt and Graupner, 1977 and Busby, 1977. The development of motorised levelling techniques is likely to continue and faster and more accurate results can be expected.

The behaviour of automatic levels was not well understood in the first few years of their use, but the following sections describe some of the work carried out in the investigation of instrumental errors since that time. Most of the systematic errors present in the early work are now reduced to very small random errors by observing techniques and the economic advantages offered have resulted in an increasing use of this equipment for all types of levelling.

10.1 Over- and under-compensation

In section 7.10 it is stated that for effective compensation at infinity focus, the enlargement factor, n is given by $n = f/s$ where f is the focal length of the objective lens system at infinity and s is the distance of the compensator from the reticule. The compensation is therefore dependent upon the accuracy with which the compensator can be positioned in the telescope. The difficulties in positioning the compensator are described by Drodofsky, 1951. The resultant positional error of the compensator will give rise to an error in compensation. Figure 10.1 shows the equivalent optical system (developed in section 7.10.2) where the compensation is in error by an amount $k\alpha$ where α is the residual inclination of sight.

Let $OR = f$,

$\quad CR = s = CR'$ and

$\quad \phi = \alpha + k\alpha$

Then $\qquad R'r = s \sin \beta = [f - (rR)] \tan \phi$

therefore, $\qquad s \sin \beta = [f - s(1 - \cos \beta)] \tan \phi$

But $\quad \beta = n\alpha$

and $\quad \phi = \alpha(1 + k)$

therefore $\quad s \sin n\alpha = [f - s(1 - \cos n\alpha)] \tan [\alpha(1 + k)]$

because α is small: $sn\alpha \simeq [f - s + s]\, \alpha\, (1 + k)$

therefore
$$1 + k \simeq \frac{sn}{f}$$

therefore
$$k \simeq \frac{sn - f}{f}$$

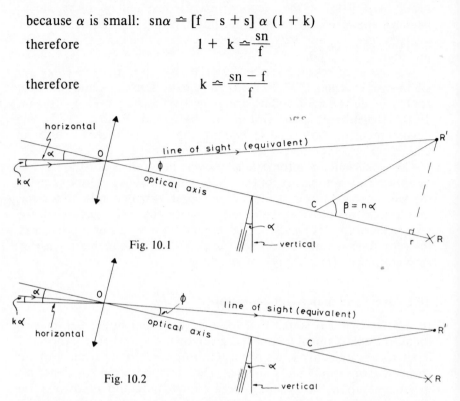

Fig. 10.1

Fig. 10.2

Compensation is correct if $k = 0$, i.e. if $n = f/s$, as before. If however, the error in positioning the compensator results in k being positive, the level is over-compensated. This is illustrated in Fig. 10.1. If the compensator is incorrectly positioned with the result that k is negative, the level is under-compensated. This is illustrated in Fig 10.2. If the objective were tilted downwards, over-compensation would result in the line of sight being inclined above the horizontal whereas under-compensation would result in the line of sight being inclined below the horizontal.

10.1.1 Compensation characteristics

Figure 10.3 illustrates the relation between residual inclination (α) of the standing axis and compensation error ($\epsilon = k\alpha$) for a hypothetical level. The range of the compensator is from $-\alpha_0$ to $+\alpha_0$ and it can be seen that the level is under-compensated throughout its range except when $\alpha = 0$. Also, k is independent of α, because the graph is a straight line. Figure 10.4 illustrates the characteristic of an over-compensated level.

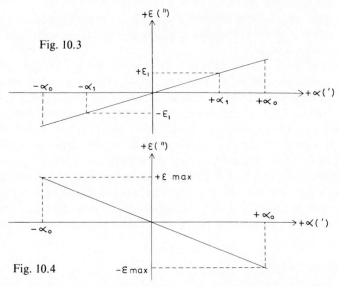

Fig. 10.3

Fig. 10.4

Suppose a level with a characteristic similar to that in Fig 10.3 is set up with a residual inclination α_1 of the standing axis and pointed towards the backstaff (α_1 assumed positive). The line of sight will be inclined upwards at an angle ϵ_1 to the horizontal (Fig 10.5a). If the backstaff is at a distance D from the level, the staff reading will be in error by approximately $+ D\epsilon_1$. If now the telescope is rotated 180° about the standing axis so that it is directed towards the forestaff, the line of sight will be inclined at an angle ϵ_1 downwards (Fig. 10.5b) so that the staff reading will be in error by $-D\epsilon_1$ assuming equal backsight and foresight distances. Thus the difference in height from

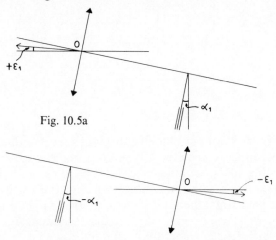

Fig. 10.5a

Fig. 10.5b

the two readings will be in error by $2\epsilon_1$. It is shown in section 10.2 that this error becomes systematic for a line of levels if the spherical bubble is out of adjustment.

If this error is to be eliminated at each set-up of the level, the characteristic should be symmetrical about the ϵ-axis (Fig 10.6). This figure shows a variation in the slope of the curve which indicates that k varies with α. With such a level, errors on backstaff and forestaff are equal in magnitude and sign and hence cancel when differences are taken. If a collimation error (c) is present (arising from an inclination of the line of collimation to the optical axis) the curve will be displaced and will lie as shown by the upper curve in Fig 10.6. Adjustment of the reticule following the two-peg test enables the actual (upper) curve to be brought through the origin, although if backsight and foresight distances are equal, collimation errors will also cancel.

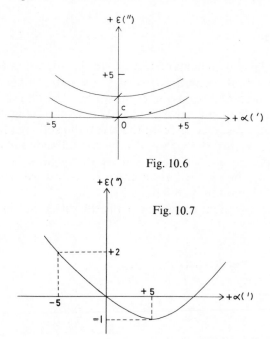

Fig. 10.6

Fig. 10.7

In practice, it is not possible to construct a level with a permanently ideal compensation characteristic, so an actual characteristic could be as shown in Fig 10.7 (assuming no collimation error). In this case, if the standing axis is set at an inclination of 5′ to the vertical, ($\alpha = +5'$ towards the backstaff) then the backsight error will be $-D \times 1/\rho''$ and the foresight error will be $+D \times 2/\rho''$ with a resultant error of $-3D/\rho''$ in the height difference. Generally, the standing axis is set vertical to much better than $\pm 5'$ so the error will be correspondingly smaller.

Obviously, the setting accuracy and adjustment of the circular level is important (section 10.2).

10.1.2 Compensation and sighting distance

The treatment so far has ignored the fact that in levelling, one is concerned with relative heights of points and not with the horizontality of the line of sight alone. Figure 7.6 illustrates the fact that the staff reading required of the compensation is that which lies on a horizontal line through the anallactic point of the telescope, where the anallactic point ideally lies on the optical axis at its intesection with the standing axis. In section 2.1.9 it is shown that it is impossible (given the physical nature of the telescope) to refer distance measurements to this fixed point and that the anallactic point changes with sighting distance, but, by suitable design, can be made almost constant for normal sighting distances. Berthon Jones, 1964 derives an expression for the error (e) in a staff reading as a function of sighting distance (D) and shows that this is closely related to the expression for the error in the multiplying constant of the telescope as a function of D. Thus, just as design of the optical components can be made to give perfect anallactic properties for a given sighting distance and optimum anallactic properties for normal sighting distances, so suitable design can give perfect compensation for one sighting distance, and optimum compensation for normal sighting distances.

The relation between the error in the staff reading (e) and the sighting distance (D) can be written as

$$\frac{e}{\sin \alpha} = kD - \lambda_1 - \frac{\lambda_2}{D}$$

where α is the inclination of the standing axis to the vertical, k is the compensation error defined in section 10.1 and λ_1 and λ_2 are constants depending on the position of the standing axis and the geometrical optics of the lenses in the telescope respectively.

An additional requirement is for the line of collimation and the optical axis to coincide when the standing axis is vertical and for both then to be horizontal. An error (c) in the collimation (i.e. an inclination c of the line of collimation to the optical axis) can be included in the above equation so that an additional error of magnitude $e_1 - D \tan c$ occurs in the staff reading.

Thus the total error (E) in staff reading is given by

$$E = e_1 + e \simeq Dc + \alpha \left[kD - \lambda_1 - \frac{\lambda_2}{D} \right]$$

Berthon Jones, 1964 followed this theoretical treatment by tests on several instruments to determine values of e for varying inclinations at

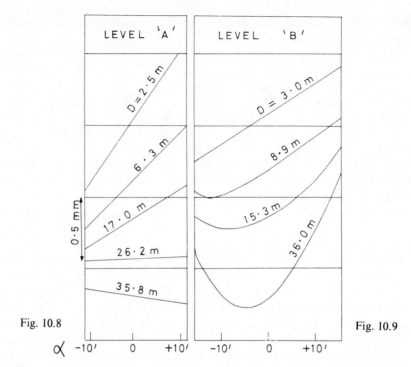

Fig. 10.8   Fig. 10.9

different sighting distances D. α is considered positive when the objective is tilted upwards and negative when tilted downwards. Figure 10.8 and 10.9 illustrate results of tests on two levels, both of which make use of a suspended prism. The ordinates represent the change in staff reading arising from a tilt, α, of the standing axis.

The following conclusions can be drawn from these results:

1. Level 'A' is under-compensated at short focus, but for sighting distances of about 30 m it is almost completely compensated. At longer focus it is over-compensated.
2. The amount of under- or over-compensation is independent of α for level 'A' for a given sighting distance, but k depends on α for level 'B'.

Other levels of the same types will not necessarily display the same characteristics because the graphs in Figs 10.8 and 10.9 depend upon the adjustment of each particular compensator. On the other hand, the equation

$$e = \alpha \left(kD - \lambda_1 - \frac{\lambda_2}{D} \right)$$

can be rearranged as: $\dfrac{e}{\alpha D} = k - \dfrac{\lambda_1}{D} - \dfrac{\lambda_2}{D^2}$

and if $e/\alpha D$ is plotted against $1/D$ then the instruments of the same type exhibit similar curves; different adjustments (and hence different values of k) simply displace the particular curve along the ordinate axis.

Berthon Jones, 1964 gives results of tests on a number of levels and shows compensation characteristics for different types of levels.

10.2 The effect of an error in the spherical level

Ideally, the standing axis should be vertical when the spherical bubble is central. If the principal tangent plane of the vial is perpendicular to the standing axis, then centring the bubble will ensure the standing axis is vertical. However, if there is an inclination between the axis and the principal tangent plane, the standing axis will have a residual inclination α when the bubble is central.

If during series levelling, the level is set up and the unadjusted spherical bubble centred with the telescope pointing always towards the backstaff (or always towards the forestaff) then the residual inclination, α, of the standing axis in the direction of the line of sight will always bear the same relation to the vertical and the direction of levelling. If such a level also exhibits an asymmetrical compensation characteristic (such as that illustrated in Fig 10.3) then a systematic error will develop along the line.

Suppose the spherical level error is such that the objective is always elevated when the bubble is central, and that the residual inclination is $+ \alpha_1$. Suppose further that the compensation characteristic is as illustrated in Fig 10.3. If the level is set up with the objective pointing towards the backstaff (at a distance D) the reading will be in error by $+D\epsilon_1$. On rotating the telescope 180° to sight the forestaff (also at distance D) the residual inclination will be $-\alpha_1$ and the reading will be in error by $-D\epsilon_1$ giving a total error of $2D\epsilon_1$. If the process is repeated at the next set-up, the error will be similar and therefore systematic. This systematic error is often referred to as an 'obliquity of the horizon' because readings are in fact referred to a datum plane inclined at an angle ϵ_1 to the horizontal. This situation is illustrated in Fig 10.10.

The reader should note that this error becomes systematic only if both a spherical bubble error and a non-symmetrical compensation characteristic are present. Another point to note is that if α is of the order of a few minutes, ϵ is of the order of one or two seconds.

In precise levelling, it is necessary to take account of this systematic error. It is possible to reduce it to a random error over a pair of set-ups by always centring the spherical bubble with the telescope pointing towards the same staff. This fits in very well with the field technique described in section 9.6.3 to reduce the effects of a change in

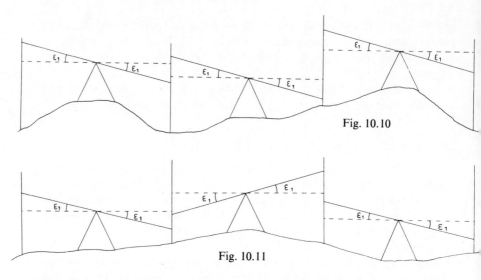

Fig. 10.10

Fig. 10.11

instrument height. If one staff of the pair is always read first, the direction of the horizon obliquity is reversed at alternate set-ups (Fig 10.11).

Drodofsky, 1957 first pointed out the existence of this error and Förstner, 1959 proposed the above observing technique. The sequence of readings in precise levelling should be therefore BFFB, FBBF, BFFB etc. where B = backstaff readings and F = forestaff readings.

Berthon Jones, 1964 describes an alternative method for the reduction of this horizon obliquity, attributed to Drodofsky, 1957. This method consists of setting up the level at each set-up so that two footscrews are normal to the direction of the second sight. After taking the readings to the first staff, the telescope is directed towards the second staff and the bubble re-centred using the third footscrew. Thus the effect of horizon obliquity is made a random error at each set-up, but an additional error is introduced arising from a change in the height of the anallactic point after re-levelling. If the observing procedure is BFFB, FBBF, etc. then this new error is random. Which of the two methods is theoretically the better depends upon the likely magnitude of each error. Berthon Jones, 1964 shows that the errors arising from a change in the height of the anallactic point are likely to be less than those arising from random horizon obliquity (except for 'periscope' levels such as the Zeiss (Jena) Ni 007).

Thus the 'relevelling' technique is generally preferable on theoretical grounds, but the simpler technique is preferable on practical grounds; the advantages offered by automatic compensation are reduced if the bubble has to be re-centred half-way through the observations at each set-up.

10.3 The effect of lateral tilts

As a result of the inevitable residual errors of construction, there will be an obliquity between the line of collimation of a telescope and the plane of symmetry of the compensator. Berthon Jones, 1964 tested various levels for this obliquity and found it to be of the order of a few minutes in a well-adjusted level. The affect of such an obliquity on the line of collimation when the standing axis is tilted in various directions relative to the line of sight is to tilt it through about 0.1″ (for a lateral tilt of the standing axis of 1′).

The test for obliquity of the plane of symmetry of the compensator is a laboratory test and if it is found to be greater than about 5′ it should be adjusted before using the level for very precise work. The adjustment should be carried out by the manufacturer.

10.4 The effect of mechanical hysteresis (compensator drag)

If a compensator with a characteristic similar to that shown in Fig 10.6 is taken from one end of the operating range (say − 10′) through zero to the other end (say + 10′) and then back through zero to − 10′, the characteristic for the − 10′ to + 10′ movement is unlikely to be identical to the + 10′ to − 10′ movement owing to hysteresis effects in the suspension. Thus the characteristic could be as illustrated in Fig 10.12. The possibility of this error was first mentioned by Drodofsky, 1957. Thus the error in compensation (ϵ) depends upon the direction from which the compensator has come.

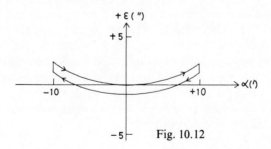

Fig. 10.12

If the compensator is between the standing axis and the reticule, rotation of the telescope about the standing axis will result in centrifugal force deflecting the compensator towards the eyepiece. Thus for the second reading at each set-up, the compensator will have come from the eyepiece. If the effect of hysteresis is to be eliminated the compensator for the first reading must also reach equilibrium by moving from the eyepiece end of the telescope. Drodofsky, 1957 pointed out that this can be achieved if, after the initial levelling-up with the telescope pointing towards the correct staff, it is rotated 360° about the standing axis before the reading is taken.

Berthon Jones, 1964 points out the possible unsatisfactory nature of the method; there is no certainty that the point from which the compensator returns to equilibrium is always the same, nor is it certain that the same compensation characteristic curve is always followed. In order to decide on these problems and on the magnitude of the hysteresis error, tests on several levels were carried out.

These showed that rapid levelling-up is preferable to slow levelling-up and that the magnitude of the errors is about 0.3″, but the reader should note that it cancels with equal staff distances if the recommended procedure is followed. The random errors were found to be about ± 0.5″.

The method for eliminating horizon obliquity by re-levelling between readings (section 10.2) also reduces the hysteresis error to a random error.

Another result of the investigation was that 'tapping' the compensator sometimes makes no difference and sometimes increases the effect of hysteresis. Tapping is not recommended; if a compensator does stick against a stop, it should be repaired by the manufacturer.

10.5 The effects of temperature

Evidence of a direct correlation between temperature and compensator error is not conclusive. Ochsenhirt, 1956 found a correlation but Neubert and Werrmann, 1963 found only a partial regular relationship. In the former case, heating was artificial, by means of lamps and in the second case, the investigation was made under normal field conditions.

In the absence of any other evidence, it seems probable that errors from temperature variations in automatic levels are less than those in instruments with spirit levels. Shielding the level from direct sunlight by an umbrella is however still desirable in high-order work; the tripod will be affected by direct sunlight.

10.6 Accuracy of the spherical spirit level

It should already be clear from the preceding sections in this chapter that it is desirable to have a well-adjusted spherical spirit level. These generally have a sensitivity of 10′ per 2 mm. If the bubble is centred carefully the standing axis can be set within about ± 20″ of the vertical.

The adjustment of the spherical level is described in section 9.1.1.

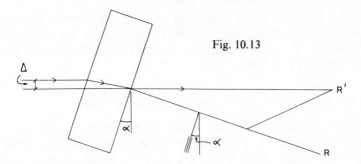

Fig. 10.13

10.7 Effect of inclined parallel plate micrometer

If the standing axis is inclined as shown in Fig. 10.13 so that the objective is raised, the parallel plate will be inclined at an angle α to the vertical. Thus the error in staff reading, $\triangle$ (assuming perfect compensation) is given by

$$\triangle \simeq t\,\alpha\,\left(1 - \frac{1}{\mu}\right)$$

where t is the thickness of the plate and μ its refractive index.

The inclination α is not likely to be greater than 20″, so

$$\triangle < t\,\alpha\,\left(1 - \frac{1}{\mu}\right)$$

Taking t = 15 mm and μ = 1.5,

$$\triangle < \frac{15 \times 20/\rho''}{3} \text{ mm}$$

i.e. $\triangle < 0.0005$ mm

On rotating the telescope 180° about the standing axis to the foresight (Fig 10.14), the error in staff reading will be $-\triangle$. Thus for one set-up, the error in height difference will be about 0.001 mm. If there is

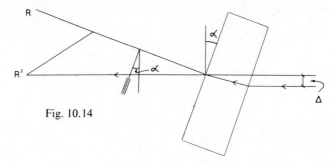

Fig. 10.14

an error in the spherical level, this will become a systematic error. The error can be made random over one set-up by the re-levelling technique or it can be made random over a pair of successive set-ups by observing BFFB, FBBF, etc.

10.8   Errors caused by external magnetic fields

Recent laboratory work by Rumpf and Meurisch, 1981 has indicated that even the relatively weak magnetic field of the earth can have a significant systematic effect on geodetic levelling carried out with levels having mechanical compensators. The effect has been shown to be of the order of 20 mm over 20 km in the case of one line of geodetic levelling in West Germany.

Similarly, precise levelling with automatic levels in workshops where large external magnetic fields are present (from electromagnets for example) is also likely to be affected.

It seems as if laboratory calibration could allow the systematic effects to be removed from the measurements, but suitable methods for doing this have not yet been fully developed.

10.9 Errors in the rotating laser level

There are two main errors to consider. Firstly, the axis of rotation may not be vertical and secondly, the laser beam may not be normal to the

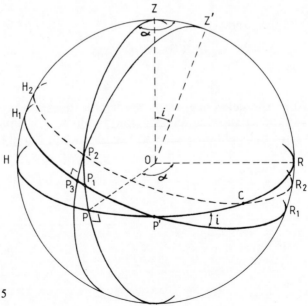

Fig. 10.15

axis of rotation. To consider the effects of these errors, it is convenient to use spherical trigonometry according to the principles outlined in section 4.7.

In Fig 10.15, the level is assumed to lie at O, the centre of a sphere. OZ is the vertical at O and HPP'R is a horizontal circle formed by the intersection of the horizontal plane through O with the sphere. If the level is in perfect adjustment, the axis of rotation is along OZ and the laser sweeps out the horizontal plane HPP'R. Now assume that the axis of rotation is inclined by an angle i to the vertical and is along OZ'. The laser sweeps out the plane $H_1P_1P'R_1$, inclined at i to the horizontal. If a staff is held vertically at P, the laser beam will intersect the staff at P_1 instead of at P, and the reading will be in error by an amount $e_1 = PP_1$, remembering that e_1 is expressed in angular measure.

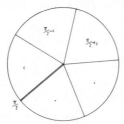

Fig. 10.16a

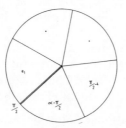

Fig. 10.16b

Obviously, the magnitude and sign of the angular error e_1 will depend upon the direction of the staff in relation to the direction of tilt of the axis of rotation of the laser. In Fig 10.15, the staff is shown at an azimuth α from this direction of tilt. If the staff were held at P' at an azimuth of $\pi/2$ from the direction of tilt, the error would be zero.

For the right-angled spherical triangle P_1PP' represented by Fig 10.16a,

$$\sin (\alpha - \pi/2) = \tan e_1 \tan (1/2 \pi - i)$$

therefore,

$$-\cos \alpha = \tan e_1 \cot i$$

Because i and e are small, this can be rearranged to give

$$e_1'' \simeq -i'' \cos\alpha$$

to first order.

If the laser beam is not normal to the axis of rotation OZ', it will sweep out the surface of a cone which will intersect the sphere in the small circle $H_2P_2R_2$. The plane of this circle is parallel to the plane of the great circle $H_1P_1R_1$. This error is analogous to a collimation error of a theodolite telescope. Suppose that the inclination of the laser beam to the normal to the axis of rotation is denoted by $c = P_2P_3$ in Fig 10.15, c being positive when the beam is inclined upwards. The additional angular error at the staff at P arising from the collimation error c is $e_2 = P_1P_2$. From the right-angled spherical triangle $P_2P_3P_1$ represented by Fig 10.16b,

$$\sin (1/2 \pi - i) = \tan c \tan (1/2 \pi - e_2)$$

therefore,

$$\cos i = \tan c \cot e_2$$

and

$$\tan e_2 = \tan c \sec i$$

But, because i, c and e_2 are all small, this equation can be rearranged to give $e_2'' \simeq c''$, to first order. Thus, the nett angular error at an azimuth α from the direction of tilt of the axis of rotation is

$$e'' = e_2'' + e_1'' \simeq c'' - i'' \cos \alpha$$

This error is zero when both c and i are zero and when $\alpha = \cos^{-1}(c/i)$. If $i > c$, there will be two positions around the horizon where the staff reading is correct (one of these positions is at C in Fig 10.15). If $i < c$, there will be no correct staff reading.

A test which will show whether significant errors c and i exist in a particular instrument can be carried out as follows. Generally, if such an error is indicated, the instrument cannot be adjusted by the user.

1. Set out two staff positions (A and B shown in plan in Fig 10.17a to e) on uniform level ground about 75 m apart.
2. Set up the laser level close to one of the staffs (A in Fig 10.17a) so that its carrying handle is approximately perpendicular to the line AB. Take the staff readings A_1 and B_1 to A and B respectively and deduce the height difference $\triangle_1 = A_1 - B_1$.
3. Turn the level anticlockwise through 90° on its base until the handle is parallel to the line AB (Fig 10.17b). Take staff readings A_2 and B_2 and deduce the height difference $\triangle_2 = A_2 - B_2$.
4. Turn the level through a further 90° (Fig 10.17c), take staff readings A_3 and B_3 and deduce $\triangle_3 = A_3 - B_3$.

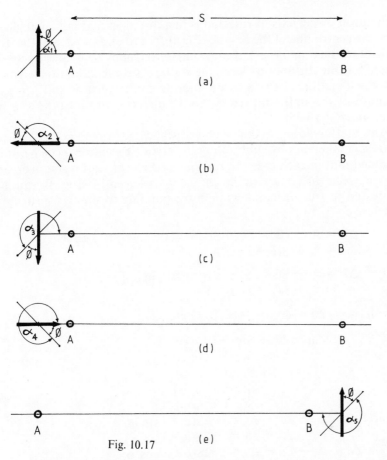

Fig. 10.17

5. Turn the level through a further 90° (Fig 10.17d), take staff readings A_4 and B_4 and deduce $\triangle_4 = A_4 - B_4$.
6. Set up the level close to the staff at B and with the carrying handle in the same direction as in step 1 (Fig 10.17e). Take staff readings A_5 and B_5 to A and B respectively and deduce $\triangle_5 = A_5 - B_5$.
7. Calculate: $2\triangle_1 - \triangle_3 - \triangle_5$,
   $$\triangle_3 - \triangle_5,$$
   $$\triangle_1 + \triangle_2 - \triangle_4 - \triangle_5 \text{ and}$$
   $$\triangle_1 - \triangle_2 + \triangle_4 - \triangle_5.$$
8. Each of these sums should be less than 10 mm for an inter-staff distance of 75 m. If one or more of the sums is greater than 10 mm, the level is in need of adjustment, although the test should be repeated with a different pair of staff positions to ensure that no gross error has been made in reading at the staffs.

The theory underlying the test is as follows. Suppose that the inclination of the axis of rotation is in a direction making an angle ϕ with the centre-line of the handle (Fig 10.17a to e). For each of the five pairs of readings in the test, the far staff is in error by $Se = S(c - i \cos \alpha)$ where S is the distance between the staffs, c is the 'collimation' error, i is the inclination of the axis of rotation to the vertical and α is the azimuth of the staff with respect to the direction of tilt of the axis of rotation (Fig 10.15).

Any staff reading taken over a distance S can be corrected for instrumental errors by subtracting the error $Se = S(c - i \cos \alpha)$. It is assumed that in each case the reading to the near staff is not in error. The correct difference in height (Δh) between A and B can be expressed in five different ways corresponding to the five measured differences Δ_1 to Δ_5:

$$\Delta h = A_j - (B_j - Sc + Si \cos \alpha_j) \text{ where } j = 1 \text{ to } 4$$
$$= \Delta_j + S(c - i \cos \alpha_j)$$

and
$$\Delta h = A_j - Sc + Si \cos \alpha_j - B_j \text{ where } j = 5$$
$$= \Delta_j - S(c - i \cos \alpha_j)$$

But, $\alpha_1 = (1/2\pi - \phi)$, $\alpha_2 = (\pi - \phi)$ etc., so

$$\Delta h = \Delta_1 + S(c - i \sin \phi)$$
$$= \Delta_2 + S(c + i \cos \phi)$$
$$= \Delta_3 + S(c + i \sin \phi)$$
$$= \Delta_4 + S(c - i \cos \phi)$$
$$= \Delta_5 - S(c + i \sin \phi)$$

Therefore,
$$2\Delta_1 - \Delta_3 - \Delta_5 = -2S(c - i \sin \phi)$$
$$\Delta_3 - \Delta_5 = -2S(c + i \sin \phi)$$
$$\Delta_1 + \Delta_2 - \Delta_4 - \Delta_5 = -2S(c + i \cos \phi)$$
$$\Delta_1 - \Delta_2 + \Delta_4 - \Delta_5 = -2S(c - i \cos \phi)$$

If each of these sums is less than 10 mm for $S = 75$ m, the terms $(c \pm i \sin \phi)$ and $(c \pm i \cos \phi)$ are all less than about $13''$ and the combination of collimation and dislevelment errors at all azimuths will not exceed this value. The manufacturers of the Spectra-Physics 944 Laser Level SL claim an accuracy of '0.005 ft per 100 ft' which is about $10''$.

Refraction errors will affect the readings but these are not as important as in the tests and adjustment of geodetic levels. If refraction effects are to be allowed for, equations similar to those given in section 9.5.1 can be used to correct the readings.

The five equations in Δh given above can be used to solve for the errors c, i and ϕ and the correct height difference Δh if required.

10.8   Summary

Source of error	Section	Effect	Action taken
1. Compensator error	10.1	Line of sight inclined by $\epsilon = k\alpha$	(a) Reduce α by accurate setting of the spherical level (b) Suitable design of characteristic & k.
2. Variation of compensation with distance sighted	10.1.2	Error, e, in staff reading given by $$\frac{e}{\sin \alpha} = kD - \lambda_1 - \frac{\lambda_2}{D}$$	(a) Suitable design of telescope components (b) Equal backsight and foresight distances.
3. Error in spherical level + compensator error	10.2	Horizon obliquity $\epsilon_1 = k\alpha_1$ (systematic)	(a) Reduce error in spherical level, and either (b) Always read same staff first (BFFB, FBBF etc.) or (c) Relevel between backsight and foresight
4. Lateral tilts	10.3	Line of sight inclined by $e \simeq 0.1''$ for a $1'$ lateral tilt & $5'$ obliquity	(a) Insignificant error if $\alpha < 1'$ (accurate centring of spherical bubble) (b) Adjustment if obliquity if $> 5'$
5. Hysteresis	10.4	Line of sight inclined by $e \simeq 0''.3$	(a) Rotate $360°$ after levelling up and equalise backsight and foresight distances
6. Temperature changes	10.5	Small effects on instrument, larger effects on tripod	Shield from direct solar radiation
7. Inclined parallel plate + spherical level error	10.7	Error of about 0.001 mm in Δh over one set-up	(a) Reduce error in spherical level and either (b) Relevel between B & F or (c) Observe BFFB, FBBF etc.
8. External magnetic fields	10.8	Of the order of 1 mm per km, or higher for fields of greater intensity than that of the earth	Laboratory calibration?

References

AESCHLIMANN, H., 1979. *An Instrumental System with Automatic Recording of Measurements.* Kern & Co AG, Aarau, Switzerland.

ALLAN, A. L., HOLLWEY, J. R. & MAYNES, J., 1968. *Practical Field Surveying and Computations.* Heinemann, London.

ALLAN, A. L., 1977. A note on centring an instrument. *Survey Review* XXV(198):360–367.

ANGUS-LEPPAN, P. V., 1968. Surface effects on refraction in precise levelling. *Proceedings, Conference on Refraction Effects in Geodesy and Electronic Distance Measurement*, University of New South Wales, Australia.

ANGUS-LEPPAN, P. V., 1971. The three-peg adjustment for levelling instruments. *Survey Review* XXI(159):9–15.

BALDINI, A. A., 1975. Determination of level sensitivity without moving the level from the instrument. *Survey Review* XXIII(175):3–16.

BECKER, J.-M., 1977. Experiences using motorized levelling techniques in Sweden. *Proceedings, XVe Congrès, Fédération Internationale des Géomètres*, Stockholm.

BELSHAW, G., 1959. Soil rebound in precise levelling. *Cartography* 1959/60(3):193–195.

BENNETT, G. G. & GROENHOUT, K. I., 1976. Practical theodolite levelling procedures. *The Australian Surveyor* 28(4):226–228.

BERTHON JONES, P., 1963. An investigation of systems of constrained centring. *Survey Review* XVII(127):22–34.

BERTHON JONES, P., 1964. An investigation of the instrumental sources of error in levelling of high precision by means of automatic levels. *Survey Review* XVII(132)276–286, (133):313–322, (134):346–354.

BURNSIDE, C. D., 1982. *Electromagnetic Distance Measurement.* 2nd edition. Granada, London.

BUSBY, J. R., 1977. The progress of precise levelling in New Zealand. *Surveying News* FIG Special Issue:31–36.

CLARK, B. A. J., 1967. An optical plummet. *The Journal of Scientific Instruments* (44):744 *et seq.*

COOPER, M. A. R., 1974. *Fundamentals of Survey Measurement and Analysis.* Crosby Lockwood Staples (Granada), London.

DEGENHARDT, K. & GRAUPNER, C., 1977. First experiences of using the method of motorised levelling in conjunction with the Ni 002 automatic geodetic level of VEB Carl Zeiss Jena. *Surveying News* FIG Special Issue:24–27.

DRODOFSKY, M., 1951. Neue Nivellierinstrumente. *Zeitschrift für Vermessungswesen* 76:225–231.

DRODOFSKY, M., 1956. Libellen mit Anzeige durch Glasblasen. *Deutsche Geodätische Kommission Series C* (7).

DRODOFSKY, M., 1957. Präzisionsnivellements mit Zeiss Ni 2. *Zeitschrift für Vermessungswesen* 82:430–434.

DYER, D. A., 1958. An error in the optical micrometer of a theodolite. *Empire Survey Review* XIV(107):213–219.

ELLENBERGER, H., 1957. The Ertel levels with automatic adjustment of the line of sight. *Zeitschrift für Vermessungswesen* 82.

FOLLONI, G., 1965. Study of Salmoiraghi self-indexing theodolite model 4200. *Bolletino di Geodesia e Scienze Affini* 1.

FÖRSTNER, G., 1953. Wirtschaftliches Nivellieren. *Allgemeine Vermessungsnachrichten* (7):225–231.

FÖRSTNER, G., 1959. Precision levelling with the automatic level Zeiss Ni 2 – survey and evaluation. *Zeiss Werkzeitschrift* 31:12 *et seq.*

GORHAM, B. J., 1976. Electronic angle measurement using coded elements, with special reference to the Zeiss Reg Elta 14. *Survey Review* XXIII(180):271–279.

GORT, A. F., 1980. A fully integrated, microprocessor-controlled total station. *Hewlett-Packard Journal* 31(9):3–11.

HALLER, R., 1963. Theodolite axis systems, their design, manufacture and precision. *Surveying and Mapping* 23(4).

HEUVELINK, H. J., 1925. Die Prüfungen der Kreisteilungen von Theodoliten und Universalinstrumenten. *Zeitschrift für Instrumentenkunden* 45:70–84.

HÜTHER, G., 1973. The new automatic geodetic level Ni 002 of VEB Carl Zeiss Jena. *Jena Review* 18:56–60.

JACKSON, J., 1980. Using an automatic index to correct for theodolite dislevelment. *Survey Review* XXV(198):360–367.

JACKSON, J. E., 1975. Telescopes anallactic or otherwise. *Survey Review* XXIII(176):51–58.

KERSCHNER, R. K., 1980. Mechanical design constraints for a total station. *Hewlett-Packard Journal* 31(9):12–14.

KISSAM, P., 1963. The Princeton standard test for estimating the accuracy of a level. *Surveying and Mapping* XXIII(1):61–67.

LENOX-CONYNGHAM, G. P., 1942. Italian transliteration of Greek. *Empire Survey Review* VI(43):318.

LEVI, L., 1968. *Aplied Optics, vol. 1*. Wiley, New York.

LEVI, L., 1980. *Applied Optics, vol. 2*. Wiley, New York.

MAURER, W., 1981. *Ein neues Verfahren zur Untersuchung von Theodolitteilkreisen*. International Congress of Surveyors (FIG) Commission 5, Montreux, Switzerland. August 1981.

MOORE, C. E. & SIMS, D. J., 1980. A compact optical system for portable distance and angle measurements. *Hewlett-Packard Journal* 31(9):14–15.

NEUBERT, K. & WERRMAN, W., 1963. Testing the functional efficiency of the Koni 007. *Vermessungs Informationen* 15:4–12.

OCHSENHIRT, H., 1956. Untersuchung des Zeiss-Nivelliers Ni 2 mit automatischer Horizontierung der Zeilachse. *Zeitschrift für Vermessungswesen* 81:348–353, 372–378.

OLLIVIER, F., 1963. *Instruments Topographiques*. Editions Eyrolles, Paris.

PESCHEL, H., 1977. Motorised geodetic levelling – an efficient method for precise height determination. *Surveying News*, FIG Special Issue:11–19.

RANNIE, J. L. & DENNIS, W. M., 1934 & 1936. Improving the performance of primary triangulation theodolites as a result of laboratory tests. *Canadian Journal of Research* 10(3) & 14:93–114.

RAWLINSON, C., 1976. Automatic angle measurement in the Aga 700 Geodimeter – principles and accuracies. *Survey Review* XXIII (180):249–270.

RICHESON, A. W., 1966. *English Land Measuring to 1800: Instruments and Practices.* The Massachusetts Institute of Technology Press, Cambridge, Mass. and London.

ROSS, D. A., 1979. *Optoelectronic Devices and Optical Imaging Techniques.* Macmillan, London.

ROY, S. P., 1979. The design of anallactic surveying telescopes with a positive focusing lens. *Survey Review* XXV(193):130–142.

RUMPF, W. E. & MEURISCH, H., 1981. *Systematische Änderungen der Ziellinie eines Präzisionskompensator-Nivelliers (insbesondere des Zeiss Ni 1) durch Magnetische Gleich-und Wechselfelder.* International Congress of Surveyors (FIG) Commission 5, Montreux, Switzerland. August 1981.

SCHEUFELE, H., 1977. The reversing compensator of the Ni 002 – temperature influence and station control. *Surveying News*, FIG Special Issue: 28–30.

SELTMANN, G., 1977. Equipment for motorised geodetic levelling. *Surveying News*, FIG Special Issue: 20–23.

SMITH, J.R., 1970. *Optical Distance Measurement.* Crosby Lockwood (Granada) London.

STROSCHE, H., 1977. Motorised levelling. *Surveying News*, FIG Special Issue: 8–10.

SZYMÓNSKI, J., c.1960. Untersuchung des Horizontalkreises und des Taumelfehlers beim Sekunden-Theodolit Theo 010 aus Jena. *Vermessungs Informationen* (II):2–26.

TARCZY-HORNOCH, A., 1967. Constrained centring of geodetic instruments. *MOM Review* (1):5–9.

TODHUNTER, I. & LEATHEM, J. G., 1956. *Spherical Trigonometry.* Macmillan, London.

WATT, I. B., 1963. Tests of the Zeiss Th. 3. *Survey Review* XVII(128): 76–88.

WEISE, H., 1966. Studies on improving the efficiency and accuracy of testing graduated circles. *Vermessungs Informationen* (17):4–44.

WOLTER, J., 1963. Präzisionsnivellement mit Kompensator Nivellieren? *Zeitschrift für Vermessungswesen* 88(11):458–462.

Appendix of tables

The following tables give particulars of several levels, optical and electronic theodolites and electronic tacheometers in use. Not all the models listed are still in production, but where an obsolete model is listed, the particulars given are for the last version. The list includes instruments shown at the exhibition associated with the 16th Congress of the International Federation of Surveyors (FIG) which was held in Montreux, Switzerland in August, 1981.

Table 1 Optical theodolites

Instrument	Manufacturer	Country	Telescope					H. Circle	
			Magni-fication	Obj. aper-ture (mm)	Length (mm)	Shortest focus (m)	Field of view (°)	Diam. (mm)	Grad-uation
GAVEC	Breithaupt	W. Germany	15	20	145	1.4	2.4	100	$1°/1^g$
TEKAT	Breithaupt	W. Germany	18	30	126	1.1	1.5	65	$1°/1^g$
VT-0	CTS	U.K.	15	25	127	1.8	3.0	64	$10'$
KOS	Kern	Switzerland	19	24		0.75	2.1	79	$5'/10^c$
TS-20	Leitz	U.S.A.	30	40	170	1.3	1.5	88	$1°$
TOM	Mashpriboritorg	U.S.S.R.	18	27	105	2.0	2.0	70	$10'$
NT-IS	Nikon	Japan	25	36	135	1.0	1.6	70	$1°/1^g$
TG-4d	Officine Galileo	Italy	22	28	115	1.8	1.5	49	$1°/1^g$
4138-C	Salmoiraghi	Italy	22	27	128	1.0	1.4	80	$10'$
STN-0	Slom	France	22	25	150	0.9	2.2	70	$1°$
TECOGON	Theis	W. Germany	25	35	160	1.3		94	$10'/20^c$
TO5	Wild	Switzerland	19	28		0.8	2.2	67	$5'/10^c$
Th 51	Zeiss (Ober.)	U.S.S.R.	20	30	125	1.2	2	72	$10'/10^c$
Tt	Askania	W. Germany	30	45	165	1.5	1.6	90	$1°$
Tts	Askania	W. Germany	30	45	165	1.5	1.6	90	$1°$
TEAUT	Breithaupt	W. Germany	30	40	150	1.9	1.3	72	$1°$
US 2		China	26	35	195	2.0		89	
VT-1	CTS	U.K.	25	38	137	1.8	2.0	78	$1°$
FT-1A	Fennel	W. Germany	30	40	175	1.2	1.3	90	$1°/1^g$
FT-20A	Fennel	W. Germany	30	40	175	1.2	1.3	90	$1°$
FTH-1	Fennel	W. Germany	30	40	175	1.2	1.3	90	$1°/1^g$
Eagle 60	K & E	U.S.A.	30	40	155	1.8	1.5	100	$1°$
DKM1	Kern	Switzerland	20	30	120	0.9	1.7	55	$20'/20^c$
KI-A	Kern	Switzerland	28	45	155	1.8	1.5	95	$1°/1^g$
KI-M	Kern	Switzerland	30	45		1.5	1.5	89	$1°/1^g$
KI-S	Kern	Switzerland	30	45		1.5	1.5	89	$1°/1^g$
T-60D	Leitz	U.S.A.	30	40	170	1.3	1.5	95	$1°$
TM-6	Leitz	U.S.A.	30	40	170	1.3	1.5	80	$1°$
TM-20C	Leitz	U.S.A.	30	40	170	1.3	1.5	80	$1°$
TT-4	Mashpriboritorg	U.S.S.R.	25	34	150	2.0	1.4	70	$20'$
OT Ш	Mashpriboritorg	U.S.S.R	27	40	152	2.0	1.7	95	$1°$
Te-C13	MOM	Hungary	26	45	143	2.0	1.3	77	$20'$
Te-D13	MOM	Hungary	26	45	143	2.0	1.3	77	$1°/1^g$

The accuracies and ranges quoted for EDM and the accuracies of electronic theodolites have been taken from manufacturers' publications and it is likely that the figures quoted have not all been derived under identical conditions. Nevertheless, the figures are useful for making broad comparisons between instruments.

V. Circle		Reading		Spirit level Value of 2mm run			Weight (kg)		Classification
Diam. (mm)	Grad-uation	Direct to	System	Plate (")	Altitude (")	Spherical (')	Inst.	Case	
65	1°/1ᵍ	1°/1ᵍ	Direct	40	–	10	2.5	2.2	*Lower-*
47	1°/1ᵍ	5'/5ᶜ	Opt. scale	30(t)	–	4	3.3	2.9	*order*
64	5'	1' & 5'	Vernier/Direct	45	–	20	3.0	2.3	Reading
69	5'/10ᶜ	5'/10ᶜ	Direct	–	–	2	4.3	1.8	to greater
70	1°	10'	Micro./Vernier	95	–	10	5.2	3	than 1'
70	10'	10'	Direct	45	–	10	1.9	2.0	
70	1°/1ᵍ	5'/5ᶜ	Opt. scale	60	–	10	3.1	3.0	
49	1°/1ᵍ	5'/5ᶜ	Opt. scale	50	–	–	2.7		
57	10'	5' & 10'	Vernier/Direct	30(t)	–	10	2.4	0.9	
70	1°	5'	Opt. scale	30	–	–	3.5	2.7	
75	1°	10'/20ᶜ&1°	Direct		–		2.5		
50	5'/10ᶜ	5'/10ᶜ	Direct	60	–	10	3.0	2.2	
72	10'/10ᶜ	10'/10ᶜ	Direct	45	–	–	2.2	2.9	
70	1°	20"	Opt. micro.	20	Auto.	10	4.6	2.2	*Middle-*
70	1°	1'	Opt. scale	20	Auto.	10	4.6	2.2	*order*
47	1°	20"/1ᶜ	Opt. micro.	40	Auto.	5	5.4	3.3	Reading
89		20"	Opt. scale	30	30	8	4.6	4.4	to greater
63	1°	20"	Opt. scale	45	90	17	5.2	4.7	than 1"
70	1°/1ᵍ	1'/1ᶜ	Opt. scale	40	Auto.	8	4.7	3.8	but not
90	1°	20"	Opt. scale	40	Auto.	8	4.7	3.8	greater
–	–	1'/1ᶜ	Opt. scale	40	–	8	4.3	3.8	than 1'
75	1°	1'	Opt. scale	20–30	Auto.	–	5.1		
55	20'/20ᶜ	10"/10ᶜᶜ	Opt. micro.	30	30	–	1.8	1.0	
75	1°/1ᵍ	20"/1ᶜ	Opt. micro.	30	Auto.	–	4.2	2.2	
70	1°/1ᵍ	6"/20ᶜᶜ	Opt. micro.	30	Auto.	–	4.6	2.4	
70	1°/1ᵍ	30"/1ᶜ	Opt. scale	30	Auto.	–	4.6	2.4	
90	1°	1'	Micro./Vernier	30	Auto.	10	5	3	
70	1°	6"	Micro./Vernier	30	Auto.	10	5.3	3	
70	1°	20"	Micro./Vernier	30	Auto.	10	5	3	
50	20'	10"	Opt. micro.	35	25	8	3.9	4.0	
70	1°	1'	Opt. scale	30	Auto.	10	2.9	3.0	
68	20'	20"	Opt. micro.	30	30	6	4.8	2.8	
68	1°/1ᵍ	1'/1ᶜ	Opt. micro.	30	30	6	4.8	2.9	

Table 1 (*contd*)

Instrument	Manufacturer	Country	Telescope					H. Circle	
			Magni-fication	Obj. aper-ture (mm)	Length (mm)	Shortest focus (m)	Field of view (°)	Diam. (mm)	Grad-uation
Te-D23	MOM	Hungary	26	45	143	2.0	1.3	77	$1°/1^g$
Te-D33	MOM	Hungary	26	45	143	2.0	1.3	77	$1°/1^g$
Te-D43	MOM	Hungary	26	45	143	2.0	1.3	77	$1°/1^g$
NT-1	Nikon	Japan	25	36	135	1.0	1.6	70	$1°$
NT-2A	Nikon	Japan	30	45	165	1.3	1.5	84	$1°/1^g$
NT-2S	Nikon	Japan	30	45	165	1.3	1.5	84	$1°/1^g$
NT-2E	Nikon	Japan	30	45	165	1.3	1.5	84	$1°$
NT-3A	Nikon	Japan	30	45	165	1.3	1.5	84	$1°/1^g$
TH-06D	Pentax	Japan	30	42		1.3	1.4	78	$1°/1^g$
TH-10A	Pentax	Japan	30	40	168	1.6	1.4	90	$1°/1^g$
TH-20A	Pentax	Japan	30	40	168	1.6	1.4	90	$1°/1^g$
TH-20C	Pentax	Japan	30	40	168	1.6	1.4	90	$1°/1^g$
4149-A	Salmoiraghi	Italy	30	36	172	2.0	1.4	90	$30''$
4150-NE	Salmoiraghi	Italy	30	40	172	2.4	1.5	90	$1°$
STN-27	Slom	France	27	36	150	1.0	1.7	80	1^g
TM-6	Sokkisha	Japan	30	40	170	1.3	1.5	80	$1°$
TM-10E	Sokkisha	Japan	30	45	170	1.3	1.5	80	$1°$
TM-20E	Sokkisha	Japan	30	45	160	1.3	1.5	76	$1°/1^g$
TM-20H	Sokkisha	Japan	30	45	160	1.3	1.5	76	$1°/1^g$
TS-6	Sokkisha	Japan	30	45	170	1.3	1.5	95	$1°/1^g$
TS-20A	Sokkisha	Japan	30	45	160	1.3	1.5	82	$1°/1^g$
TL-6DE	Topcon	Japan	30	42	152	1.5	1.3	70	$1°/1^g$
TL-10DE	Topcon	Japan	30	42	152	1.5	1.3	70	$1°/1^g$
TL-20E	Topcon	Japan	30	42	152	1.5	1.3	70	$1°/1^g$
TL-20p	Topcon	Japan	30	42	152	1.5	1.3	70	$1°/1^g$
TL-60SE	Topcon	Japan	30	42	152	1.5	1.3	70	$1°/1^g$
ST-456	Watts	U.K.	25	38	146	1.8	1.5	89	$20'$
TI	Wild	Switzerland	30	42	150	1.7	1.6	79	$1°/1^g$
T16	Wild	Switzerland	30	42	150	1.7	1.6	94	$1°/1^g$
Theo 020A	Zeiss (Jena)	E. Germany	25	36	180	1.5	1.3	86	$1°/1^g$
Th 1	Zeiss (Ober.)	W. Germany	30	40	155	1.6	1.4	98	$1°/1^g$
Th 42	Zeiss (Ober.)	W. Germany	30	40	155	1.6	1.4	98	$1°/1^g$
Tu	Askania	W. Germany	30	45	165	1.5	1.6	90	$20'$
FT-2	Fennel	W. Germany	30	45	174	2.0	1.6	93	$20'$
DKM2-A	Kern	Switzerland	32	45	186	1.5	1.3	75	$10'/10^c$
TM-1A	Leitz	U.S.A.	30	40	170	2.2	1.5	94	$20'$
ТБ-1	Mashpriboritorg	U.S.S.R.	26	40	180	1.2	1.3	85	$20'$
Te-B43	MOM	Hungary	30	45	143	2.0	1.3	93	$20'/20^c$
NT-5	Nikon	Japan	30	45	165	1.6	1.5	94	$10'$
TH-01W	Pentax	Japan	30	40	168	1.6	1.4	100	$20'/20^c$
4200-A	Salmoiraghi	Italy	30	40	172	2.5	1.5	90	$10'$
TM-1A	Sokkisha	Japan	30	40	170	2.2	1.5	94	$20'/20^c$
TL-1E	Topcon	Japan	30	42	152	1.5	1.3	90	$20'/20^c$
ST-400	Watts	U.K.	28	41	165	1.8	1.5	98	$10'$
T2	Wild	Switzerland	30	40	150	2.2	1.6	90	$20'/20^c$
Theo 010A	Zeiss (Jena)	E. Germany	30	40	180	2.0	1.3	86	$20'/20^c$
Th 2	Zeiss (Ober.)	W. Germany	30	40	155	1.6	1.5	100	$10'/10^c$
DKM3	Kern	Switzerland	30, 45	68	140	5	1.2	104	$10'/10^{cc}$
OT-02	Mashpriboritorg	U.S.S.R.	24, 30, 40	60	265	5.0	1.6	135	$4'$
Geod. Tavi	CTS	U.K.	20, 30	60	225	5.0	1.3	127	$20'$
Microptic 3	Watts,	U.K.	40	50	170	1.8	1.0	98	$5'$
T3	Wild	Switzerland	24, 30, 40	60	265	4.0	1.6	135	$4'/10^c$

V. Circle		Reading		Spirit level Value of 2mm run			Weight (kg)		Classification
Diam. (mm)	Grad-uation	Direct to	System	Plate (")	Altitude (")	Spherical (')	Inst.	Case	
68	$1°/1^g$	$1'/2^c$	Opt. scale	30	30	6	5	2.9	*Middle-order*
68	$1°/1^g$	$1'/2^c$	Opt. micro.	30	Auto.	6	4.8	4.1	order
68	$1°/1^g$	$1'/2^c$	Opt. scale	30	Auto.	6	5		Reading to
70	$1°$	$1'$	Opt. micro.	60	60(t)	10	3.1	3.0	greater
84	$1°/1^g$	$10''/25^{cc}$	Opt. micro.	30	Auto.	10	4.7	2.6	than 1"
84	$1°/1^g$	$1'/1^c$	Opt. scale	30	Auto.	10	4.7	2.6	but not
84	$1°$	$10''$	Coinc. micro.	30	Auto.	10	4.7	2.6	greater
84	$1°/1^g$	$5''/10^{cc}$	Coinc. micro.	30	Auto.	10	4.7	2.6	than 1'
78	$1°/1^g$	$6''/20^{cc}$	Opt. micro.	30	Auto.	8	4.0	3.5	
74	$1°/1^g$	$10''/20^{cc}$	Opt. micro.	30	Auto.	10	4.9	2.3	
74	$1°/1^g$	$20''/50^{cc}$	Opt. micro.	30	Auto.	10	4.9	2.3	
74	$1°/1^g$	$20''/50^{cc}$	Opt. micro.	60	40(t)	10	4.3	2.3	
90	$30''$	$30''$	Direct	30(t)	Auto.	10	4.7	3.1	
70	$1°$	$1'$	Opt. scale	30	30	10	4.6	2.9	
80	1^g	1^c	Opt. micro.	30	–	–	4.6	2.3	
70	$1°$	$6''$	Opt. micro.	30	Auto.	10	5.3	3.0	
70	$1°$	$10''$	Coinc. micro.	30	Auto.	10	5.6	3.0	
76	$1°/1^g$	$20''/50^{cc}$	Opt. micro.	40	40(t)	10	4.6		
76	$1°/1^g$	$20''/50^{cc}$	Opt. micro.	30	Auto.	10	4.6		
85	$1°/1^g$	$1'/1^c$	Opt. scale	30	Auto.	10	5.6		
76	$1°/1^g$	$1'/1^c$	Opt. scale	60	40(t)	10	3.9		
70	$1°/1^g$	$6''/20^{cc}$	Opt. micro.	30	Auto.	10	4	3.3	
70	$1°/1^g$	$10''/50^{cc}$	Opt. micro.	30	Auto.	10	4	3.3	
70	$1°/1^g$	$20''/1^c$	Opt. micro.	40	Auto.	10	3.3	1.5	
70	$1°/1^g$	$20''/1^c$	Opt. micro.	40	40(t)	10	3.3	1.5	
70	$1°/1^g$	$1'/2^c$	Opt. scale	40	Auto.	10	3.3	3.3	
64	$20'$	$20''$		40	30	–	5.1	3.9	
79	$1°/1^g$	$6''/20^{cc}$	Opt. micro.	30	Auto.	8	5.8	2.8	
79	$1°/1^g$	$1'/1^c$	Opt. scale	30	Auto.	8	5.3	2.8	
86	$1°/1^g$	$20''/20^{cc}$	Opt. scale	20	Auto.	8	4.2	4.4	
85	$1°/1^g$	$20''/1^c$	Opt. scale	30	–	15	4.7	4.2	
85	$1°/1^g$	$20''/1^c$	Opt. scale	30	Auto.	10	4.7	4.2	
70	$20'$	$1''$	Coinc. micro.	20	Auto.	10	4.6	2.2	*Higher-order*
60	$20'$	$1''$	Coinc. micro.	20	20	6	5.5	3.6	order
69	$10'/10^c$	$1''/1^{cc}$	Coinc. micro.	22	Auto.	–	6.2	2.4	Reading
80	$20'$	$1''$	Coinc. micro.	20	Auto.	10	6	4	to 1"
75	$20'$	$1''$	Coinc. micro.	20	25	12	5.1	3.0	
60	$20'/20^c$	$1''/2^{cc}$	Coinc. micro.	20	20	6	5.3		
74	$10'$	$1''$	Coinc. micro.	20	30	10	6.0	3.0	
80	$20'/20^c$	$1''/1^{cc}$	Coinc. micro.	20	30	10	5.5		
90	$10'$	$1''$	Coinc. micro.	20	Auto.	10	6.1	5.2	
80	$20'/20^c$	$1''/1^{cc}$	Coinc. micro.	20	Auto.	10	6.0	4.0	
70	$1°/1^g$	$1''/1^{cc}$	Coinc. micro.	20	Auto.	10	4.9	2.5	
76	$10'$	$1''$	Coinc. micro.	20	20	–	6.1	4.1	
70	$20'/20^c$	$1''/1^{cc}$	Coinc. micro.	20	Auto.	8	6.0	2.2	
86	$20'/20^c$	$1''/2^{cc}$	Coinc. micro.	20	Auto.	8	4.3	4.4	
85	$10'/10^c$	$1''/1^{cc}$	Coinc. micro.	20	Auto.	10	5.2	4.8	
104	$10'/10^{cc}$	$0''.5/0^{cc}.5$	Coinc. micro.	10	10	–	12.2	3.1	*Geodetic*
90	$8'$	$0''.2$	Coinc. micro.	7	12	–	11.0	4.2	Reading to
70	$20'$	$0''.5 \& 1''$	Coinc. micro.	20	10	–	9.8	6.3	less than 1"
76	$5'$	$0''.2$	Coinc. micro.	10	20	–	8.0	6.5	
90	$8'/20^c$	$0''.2/0^{cc}.5$	Coinc. micro.	7	13	–	11.2	3.7	

[(t) denotes level on telescope]

Table 2   Optical theodolites with integral EDM

Name	Guppy	SM41	EOT 2000
Manufacturer	Topcon	Zeiss (oberkochen)	Zeiss (Jena)
Country	Japan	W. Germany	E. Germany
Theodolite			
based on	TL series	Th 42	Theo 010A
reading to	10″	10″/1^c	1″
system	optical micrometer	optical scale	optical micrometer
EDM			
carrier wavelength	c.900 nm	910 nm	860 nm
source	GaAs diode	GaAs diode	GaAs diode
range (approx.)	1km with 3 prisms	2km with 9 prisms	2km with 7 prisms
accuracy	± (5mm + 10ppm)	± (10/20mm + 2ppm)	± 5 to 10mm
display	LEDs	Nixie tubes	LEDs
reading to	1mm	1mm	1mm
time for 1 meas.	1.6s to 5s	1s to 5s	10s
Microprocessor functions			
m/ft	yes	yes	yes
refractive index	yes	yes	yes
prism 'constant'	variable	fixed	fixed
slope reduction	no	no	yes
height difference	no	no	yes
mean FL & FR V.A.	no	no	yes
Battery	external	integral	external
Recording device	no	Rec 100 (manual)	EOT2000/3 (automatic)
Weight	6kg	6.5kg	11kg

Table 3   Electronic theodolites with modular EDM & data storage

Instrument	Vectron	E1 & E2
Manufacturer	K & E	Kern
Country	U.S.A.	Switzerland
Telescope		
magnification	30	32
aperture	40mm	45mm
shortest focus	1.8m	1.5m
field of view	1° 20′	
Angle measurement		
circle code	incremental	incremental
accuracy	± 3″	± 2″ (± 0.3″ for E2)
max. slew rate	5 rev s^{-1}	1.5 rev s^{-1}
display	8-digit gas disch.	2 × 6-digit LCD
EDM		
device	AutoRanger	DM502
range (approx.)	2km	2km with 3 reflectors
accuracy	± (5mm + 6ppm)	± (5mm + 5ppm)
reading to	1mm	1mm
time was 1 meas.	6s	8s
display		
slope dist.	8-digit gas disch.	6-digit LCD
hor. dist.	at angle display	6-digit LCD
ht. diff.	at angle display	6-digit LCD
Controlling microprocessor		
extra functions †	averages and calcn. of standard devns.	none
instructed by	hand-held keyboard, cable connection.	
Data storage		
device	Vectron field computer	R32/ R48
type	solid-state	solid-state
keyboard	alphanum. + funcs.	numeric + funcs.
display	9-digit, fully alphanum. LCD	15-digit, partially alphanum. LED
additional funcs. *	yes	no
Weight		
theodolite	6.8kg	8.7kg
EDM	2.4kg	1.6kg
storage device	1.1kg	0.5kg

† other than functions associated with registration of circle rotations, residual tilts and corrections for slope, refraction etc.

*other than functions associated with data transfer and storage.

Table 4   Electronic Tacheometers

Instrument	Geodimeter 710	3820A Elect. Total Stn.	Vectron II
Manufacturer	Aga	Hewlett-Packard	Keuffel & Esser
Country	Sweden	U.S.A.	U.S.A.
Telescope			
magnification	30	30	30
aperture		66mm	70mm
shortest focus	1.7m	5m	2m
field of view	1° 10′	1° 30′	1° 30′
Angle measurement			
circle code	incremental	absolute	incremental
accuracy	±2″(h) & ±3″(v)	±2″(h) & ±4″(v)	±1″
max. slew rate			3 rev/s
EDM			
source	HeNe gas laser	GaAs lasing diode	C.W. infra-red lasing diode
range (approx.)	5km (6 reflectors)	5km (6 reflectors)	5km
accuracy	±(5mm±1ppm)	±(5mm + 5ppm)	±(5mm + 2ppm)
reading to	1mm	1mm	1mm
time for 1 meas.	10–15s	1.5–2.5s	6s
Data display	separate, 1 × 14-digit Nixie	integral, 2 × 10-digit LED	integral, 2×40-digit LCD
Controlling microprocessor			
extra functions †	no	no, but interfacing to HP computers possible	yes
instructed by	separate keyboard, cable connection	integral keyboards, one on each face	separate keyboard, infra-red signal link
Data storage			
device	Geodat 120 + interface	HP 3851A	integral
type	solid-state, automatic	solid state, automatic	solid-state, automatic
keyboard	numeric + funcs.	numeric + funcs.	microproc. keyboard used
display	15-digit partially alphanum. LED	20-digit LED	at data displays
additional funcs. *	no	no, but interfacing to HP computer possible	no
Weight			
tacheometer	14.2kg + 10.8kg display/ control unit	9.6kg	13kg
storage device	0.6kg	0.6kg	included

† other than functions associated with registration of circle rotations, residual tilts and corrections for slope, refraction, etc.

* other than functions associated with data transfer and storage.

Table 4 *Continued*

Tachymat TCIL	Elta 2	Elta 3 & Elta 20	Elta 4
Wild	Zeiss (Oberkochen)	Zeiss (Oberkochen)	Zeiss (Oberkochen)
Switzerland	W. Germany	W. Germany	W. Germany
25	30	30	25
60mm	60mm	60mm	40mm
2m	2.5m	2.5m	3m
1° 50′			
incremental	absolute	absolute	incremental
±2″(h) & ±3″(v)	±0.2mg on (±0.6″)	±2″ (Elta 3), ±1″ (Elta 20)	±3″
2 rev/s			
GaAs LED	GaAs LED	GaAs LED	GaAs LED
5km (11 reflectors)	4km (9 reflectors)	2.5km (Elta 3), 3km (Elta 20) (9 reflectors)	2.2km (9 reflectors)
±(5mm + 5ppm)	±(5mm + 2ppm)	±(5mm + 2ppm)	±(5mm + 2ppm)
1mm	1mm	1mm	1mm
4–15s	1–5s	1–5s	1–5s
integral, 2 × 16-digit LED	integral, 2 ×9-digit LED	integral, 9-digit LED	integral, 9-digit LED
yes	Yes, considerable on-site computing using ROMs	yes	no
integral keyboards, one on each face	integral thumbswitches, 14-digits	integral thumbswitches, 14-digits	integral switches
Recording Attachment mag. tape cassettes, auto. microproc. keyboard used at data displays	MEM (integral) solid-state, auto. microproc. thumbschs. used at data displays	MEM (integral) solid state, auto. microproc. thumbsw. used at data displays	REC100 electronic fieldbook solid-state, manual numeric + funcs. 9-digit LED
no	yes, using DAC100 data converter	Yes, using DAC100 data converter	yes, using DAC100 data converter
9.8kg	13.5kg	13.5kg	6.5kg
2.1kg	included	included	0.6kg

Table 5   Non-Automatic Levels

Instrument	Manufacturer	Country	Type	Telescope					Spirit levels value of 2mm run		Weight (kg)		Classification
				Magni-fication	Obj. aper-ture (mm)	Length (mm)	Shortest focus (m)	Field of view (°)	Main (")	Spherical (")	Inst.	Case	
Lb	Askania	W. Germany	Tilting	25	32	180	1.2	1.7	20	10		2	Levels with bubbles viewed by mirrors.
NAKUM	Breithaupt	W. Germany	Dumpy	25	30	155	0.9		30	6	1.4	2.2	
NAMAL	Breithaupt	W. Germany	Tilting	30	40	190	1.2		25	6	2.1	2.5	
NIFIX	Breithaupt	W. Germany	Tilting	20	30	127	0.8		45	30	1.0	0.3	
CG 20	Clarkson	U.K.	Tilting	20	32		1.5		30	10			
VLO-C	CTS	U.K.	Tilting	19	28	151	1.8	2.5	60	8	1.3	0.7	
BNL	Ertel	W. Germany	Tilting	18	25	160	1.5	2.6	60	30	0.6	0.4	
Fen 1	Fennel	W. Germany	Tilting	25	35	160	1.3		60	–	1.3	0.8	
Ni 3L	Fennel	W. Germany	Dumpy	25	30	200	1.6		30	8	1.8	2.4	
GK 0	Kern	Switzerland	Tilting	18	24	127	0.9	2.1	50	15	0.8	0.5	
Ni-E3	MOM	Hungary	Tilting	14	30	175	3		60		1.2	0.7	
5150-A	Salmoiraghi	Italy	Dumpy	22	27	155	1.0	1.4	30		1.5	0.7	
5153-C	Salmoiraghi	Italy	Tilting	22	27	155	1.0	1.4	30	10	1.7	0.7	
SN 0	Slom	France	Tilting	22	25	150	0.9	2.0	25	10	1.2	1.0	
Sitemaster	Stanley	U.K.	Tilting	14	25	178	1.5	2.0	80	8	1.2	0.3	
N4	Theis	W. Germany	Tilting	25	35	160	1.3		30	15	1.4	0.8	
N5	Theis	W. Germany	Dumpy	25	35	160	1.3		30	15	1.4	0.8	
N6	Theis	W. Germany	Dumpy	25	35	160	1.3		60	15	1.2	0.8	
N44	Theis	W. Germany	Tilting	25	30	160	1.3		30	30	1.2	0.8	
SL 410	Watts	U.K.	Tilting	14	30	170	3.0	2.6	60	8	1.3	0.7	
N 01	Wild	Switzerland	Dumpy	19	25	160	0.8	2.3	60	8	1.7	–	
N 05	Wild	Switzerland	Tilting	19	25	160	0.8	2.3	60	8	1.7	–	
Ni 060	Zeiss (Jena)	E. Germany	Tilting	19	25	138	1.5	2.1	60	8	0.9	1.0	
Ni-3	Zeiss (Ober.)	W. Germany	Dumpy	19	25	135	1.2	2.0	30	15	1.2	1.2	
Ni-52	Zeiss (Ober.)	W. Germany	Tilting	20	25		1.0	2.0	60	30	0.8	0.7	
VL2	CTS	U.K.	Tilting	28	38	194	2	1.8	60	8	2.0	2.5	Lower-order tilting levels with coincidence bubble viewing.
KOPNO	Fennel	W. Germany	Tilting	25	30	200	1.6		30	8	2.0	2.4	
GK 1	Kern	Switzerland	Tilting	22	30	118	0.9	1.7	50	15	0.9	0.6	
HJT-3	Mashpriboritorg	U.S.S.R.	Tilting	30	40	180	2.0	1.3	30	7	2.1	2.7	
E5	Nikon	Japan	Tilting	25	40	245	1.5	1.2	40	10	2.2	2.3	

Appendix of tables 253

Model	Manufacturer	Country	Type									
L106	Officine Galileo	Italy	Tilting	22	32	170	1.5	1.7	40	10	2.6	1.5
L-30	Pentax	Japan	Tilting	32	45	197	1.5	1.2	40	10	2.4	1.6
5167	Salmoiraghi	Italy	Tilting	25	35	160	3.0	1.7	30	10	3.3	3.1
SN 1	Slom	France	Tilting	22	25	150	0.9	2.0	40	10	1.3	1.1
TTL6	Sokkisha	Japan	Tilting	25	40	240	1.8	1.3	40	10	2.2	2.5
TS 2	Topcon	Japan	Tilting	32	45	217	1.4	1.3	40	10	2.0	1.2
SL432	Watts	U.K.	Tilting	28	38	180	1.8	1.5	30	10	1.9	2.4
Ni 030	Zeiss (Jena)	E. Germany	Tilting	25	35	195	2.0	1.6	30	8	1.6	1.6
Li (†)	Askania	W. Germany	Tilting	30	45	185	1.7	1.5	20	10		2.5
NAGUK	Breithaupt	W. Germany	Tilting	30	40	190	1.2		20	6	1.9	
ND-1		China	Tilting	30	45	180	3.0		15	10	1.7	
IGN1	Fennel	W. Germany	Tilting	30	40	200	1.6		20	8	2.0	2.4
GK 23 (†)	Kern	Switzerland	Tilting	30	45	170	1.8	1.4	18	6	1.5	1.9
HB-1	Mashpriboritorg	U.S.S.R.	Tilting	31	40	175	3.0	1.6	17	7	1.8	2.0
Ni-B11 (†)	MOM	Hungary	Tilting	28	40	160	1.2	1.3	20	6	2.6	1.7
L103	Officine Galileo	Italy	Tilting	22	32	170	1.5	1.7	20	10	2.6	1.5
L-20 (†)	Pentax	Japan	Tilting	32	45	197	1.5	1.2	20	10	3.0	1.9
PL1 (†)	Sokkisha	Japan	Tilting	42	50		2	1.2	10	4	4.6	4.0
5169 (†)	Salmoiraghi	Italy	Tilting	30	45	160	4.0	1.4	20	10	3.4	3.6
SN 3C	Slom	France	Tilting	35	50	260	4.0	0.9	13	5	4.3	1.7
SN 2C	Slom	France	Tilting	27	36	150	1.0	1.7	20	10	1.9	2.3
N 2 (†)	Wild	Switzerland	Tilting	30	40	196	1.7	1.7	30	8	2.2	1.3
NABON	Breithaupt	W. Germany	Tilting	42	50	335	2.0		8	–	7.4	4.7
S500	CTS	U.K.	Tilting	36	51	298	3.0	1.5	15	20	5.4	3.7
HA-1	Mashpriboritorg	U.S.S.R.	Tilting	42	56	400	3.0	0.9	10		5.8	5.3
NB-2		U.S.S.R.	Tilting	49	60	420	4.2		10	4	6.0	6.2
Ni-A1	MOM	Hungary	Tilting	40	65	314	2.5	1.1	10	2	4.2	2.9
L-10	Pentax	Japan	Tilting	42	50		2	1.8	10	4	3.7	2.3
N 3	Wild	Switzerland	Tilting	42	52	297	0.5	1.0	10	4	5.1	3.7
Ni 004	Zeiss (Jena)	E. Germany	Tilting	44	56	375	3.0	1.0	10	25	6.1	4.2

Higher-order tilting levels with coincidence bubble viewing.

(†) denotes parallel-plate micrometer available.

Geodetic levels with parallel-plate micrometers.

Table 6 Automatic Levels

Instrument	Manufacturer	Country	Magnification	Obj. aperture (mm)	Telescope Length (mm)	Telescope Shortest focus (m)	Telescope Field of view (°)	Compensator Type	Compensator Damping	Spherical level value of 2mm run (')	Weight (kg) Inst.	Weight (kg) Case	Classification
AUTOM	Breithaupt	W. Germany	27	33	235	1.5		Suspended mirror	Magnetic	5	2.6	2.5	Lower-order.
BNA	Ertel	W. Germany	24	30	160	1.3	1.7	Supported prism	Magnetic	15	1.4	1.0	
FNA-1	Fennel	W. Germany	25	30	200	1.9		Suspended prism	Air	15			
AL-3	K & E	U.S.A.	38	24	260	2	1.3	Suspended prism	Air	10	3		
GK-0A	Kern	Switzerland	21	30	170	0.8	1.7	Suspended prism	Magnetic	20	1.9	0.8	
GK-1A	Kern	Switzerland	25	45	125	2.3	1.4	Pivoted lens	Air	15	1.6	1.1	
PAL-5C	Pentax	Japan	24	38		0.5	1.3	Pivoted prism	Air	10	1.7	1.6	
AZ-1	Nikon	Japan	22	30	205	0.8	1.5	suspended prism	Air	10	1.5	0.9	
5173	Salmoiraghi	Italy	30	30	'Periscope'	2.0	1.3	Suspended prism	Air	10	3.4	1.9	
C3A	Sokkisha	Japan	25	30	210	1.6	1.3	Suspended objective	Magnetic	10	2.2	1.4	
TECOMAT	Theis	W. Germany	25	35	220	2.2	1.8	Suspended prism	Magnetic	8	2.4	2.4	
AT-D3	Topcon	Japan	26	30	245	1.8	1.3	Suspended prism		10	2.2	1.2	
AT-D4	Topcon	Japan	26	30	241	1.0	1.3	Suspended prism		10	2.2	1.2	
AT-M3	Topcon	Japan	26	40	238	0	1.7	Suspended prism		10	2.3	1.2	
SL 305	Watts	U.K.	24	36		0.9	1.5	Suspended prism	Air	10			
NA 0	Wild	Switzerland	20	30		0.9	2.0	Suspended prism	Air	8	1.8	1.5	
NA 1	Wild	Switzerland	24	36		1.0	1.8	Suspended prism	Air		2.1	1.5	
Ni 050	Zeiss (Jena)	E. Germany	18	25	210	0.8		Suspended prism	Air	25	1.0	1.3	
Ni 025	Zeiss (Jena)	E. Germany	25	30	195	1.5	1.6	Suspended prisms	Air	8	1.7	1.7	
Ni 42	Zeiss (Ober.)	W. Germany	22	25	135	1.2	1.3	Suspended mirror	Magnetic	20	1.4	0.7	
Na 1	Askania	W. Germany	25	36	170	1.8	1.6	Supported mirror	Air	10	2.3	0.7	Higher-order.
S77 (†)	CTS	U.K.	32	40	267	1.8	1.3	Suspended prisms	Air	15	2.7	5.0	(†) denotes parallel-plate micrometer available.
INA (†)	Ertel	W. Germany	32	40	185	1.7	1.1	Supported prism	Magnetic	8	2.0	2.4	

AUTAC (†)	Fennel	W. Germany	32	40	250	2.5		Suspended prism	Air	8	3.0	2.2
Eagle 2	K & E.	U.S.A.	32	40	260	3.3	1.5	Suspended prism	Air	10	2.6	3.1
GK-2A (†)	Kern	Switzerland	32	45		2.2	1.3	Suspended prism	Air	10	3.5	3.1
Ni-B5 (†)	MOM	Hungary	32	45	272	3.0	1.3	Suspended prism	Air	8	2.3	2.3
AE-5 (†)	Nikon	Japan	28	40	214	0.5	1.5	Suspended prism	Air	10	1.6	1.5
AP-5	Nikon	Japan	26	30	205	0.8	1.5	Suspended prism	Air	10	1.5	0.9
LA 102 (†)	Officine Galileo	Italy	30	40	200	1.8	1.3	Suspended prism	Air	10	3.1	1.5
PAL-2c	Pentax	Japan	32	45		0.5	1.3	Suspended prism	Air	10	1.7	2.5
PAL-3c	Pentax	Japan	28	42		0.5	1.3	Suspended prism	Air	10	1.7	2.5
5190	Salmoiraghi	Italy	30	45	'Periscope'	3.9	1.4	Suspended reticule	Air	10	6.9	6.5
SNA-2	Slom	France	25	40	217	1.7	1.6	Suspended mirror	Magnetic	10	1.7	1.4
B1 (†)	Sokkisha	Japan	32	45	270	2.3	1.3	Suspended prism	Magnetic	10	3.0	3.0
B2A (†)	Sokkisha	Japan	30	40	210	1.8	1.3	Suspended prism	Magnetic	10	2.6	1.4
B2C (†)	Sokkisha	Japan	32	40	220	1.4	1.3	Suspended prism	Magnetic	10	2.2	1.9
TECOMAT 32	Theis	W. Germany	32	40	220	2.2	1.7	Suspended prism	Magnetic	8	2.4	2.4
AT-F (†)	Topcon	Japan	32	45	259	1.5	1.3	Suspended prism	Magnetic	10	2.1	2.8
AT-D2 (†)	Topcon	Japan	32	40	239	1.8	1.3	Suspended prism	Magnetic	10	2.3	1.2
NA 2 (†)	Wild	Switzerland	40	45	250	1.6	1.3	Suspended prism	Air	8	2.4	1.8
Ni 007 (†)	Zeiss (Jena)	E. Germany	32	40	'Periscope'	2.2	1.3	Suspended prism	Air	8	3.9	2.8
Ni 2 (†)	Zeiss (Ober.)	W. Germany	32	40	270	3.3	1.3	Suspended prism	Air	10	2.1	3.3
Ni-A31	MOM	Hungary	50	67	210	2.5	1.0	Suspended mirror	Air	8	3.0	
Ni 002	Zeiss (Jena)	E. Germany	40	55	370	1.5	1.1	Suspended mirror	Air	8	6.5	5.1
Ni 1	Zeiss (Ober.)	W. Germany	50	50	305	1.4	1.0	Suspended prism		5	5.2	4.6

Geodetic
With parallel-plate micrometer.

Index

258 *Modern theodolites and levels*